WEB WRITING/
WEB DESIGNING

WEB WRITING/
WEB DESIGNING

MARGARET W. BATSCHELET
University of Texas at San Antonio

ALLYN AND BACON

Boston ■ London ■ Toronto ■ Sydney ■ Tokyo ■ Singapore

Vice President, Humanities: *Joseph Opiela*
Series Editorial Assistant: *Julie Hallett*
Executive Marketing Manager: *Lisa Kimball*
Production Editor: *Christopher H. Rawlings*
Editorial-Production Service: *Omegatype Typography, Inc.*
Composition and Prepress Buyer: *Linda Cox*
Manufacturing Buyer: *Suzanne Lareau*
Cover Administrator: *Brian Gogolin*
Electronic Composition: *Omegatype Typography, Inc.*

Between the time Website information is gathered and then published, it is not unusual for some sites to have closed. Also, the transcription of URLs can result in unintended typographical errors. The publisher would appreciate notification where these occur so that they may be corrected in subsequent editions. Thank you.

Many of the designations used by manufacturers and sellers to distinguish their products are claimed as trademarks. Where those designations appear in this book, and Allyn and Bacon was aware of a trademark claim, the designations have been printed with an initial capital. Designations within quotation marks represent hypothetical products.

Library of Congress Cataloging-in-Publication Data

Batschelet, Margaret.
    Web writing/Web designing / Margaret Batschelet.
        p.   cm.
    Includes bibliographical references (p.).
    ISBN 0-205-31742-1
    1. Web sites—Design.   2. Technical writing.   3. HTML (Document markup language)   I. Title

    TK5105.888. B378   2001
    005.7'2—dc21

                                                    00-032276

Printed in the United States of America

10   9   8   7   6   5   4   3   2   1   05   04   03   02   01   00

Credits: Netscape Communications Corporation has not authorized, sponsored, endorsed, or approved this publication and is not responsible for its content. Netscape and the Netscape Communications Corporate Logos are trademarks and trade names of Netscape Communications Corporation. All other product names and/or logos are trademarks of their respective owners. Microsoft Screen shots reprinted by permission from Microsoft Corporation.

CONTENTS

CHAPTER THREE
GIFs and JPEGs and PNGs, Oh My: Using Images   33

CHAPTER FOUR
How Should I Say It: Writing for the Web   57

CHAPTER FIVE
Laying It All Out: Tables   73

CHAPTER SIX
How Should It Look: Web Page Design   103

CHAPTER SEVEN
Making the Web a Web: Links   123

CHAPTER EIGHT
How Should It Work: Site Structure   149

APPENDIX
Creating Frames 197

WEB WRITING

Web Writing/Web Designing is a book about writing for the World Wide Web, but we need to begin by looking at what *writing* means in this context. Web writing involves more than just words; in addition to text, Web writers use colors, images, layout, and code to produce pages that visitors to their sites will want to read. Some of the techniques that go into these pages are very creative—they involve conceptualizing the way you want a page to look, read, and "behave," and also implementing the tools that will create such a page. But some of the tasks involved are a little more daunting: writing code, saving images at the right size, and checking the way your pages appear in several browsers. Doing these things may not be as exciting as conceptualizing a page, but, like other kinds of revision and polishing, they're necessary if you want to bring your concept into existence.

The process of writing Web pages has much in common, in some ways, with the process of writing in general. Web writers, like paper writers, develop ideas and ways to make those ideas come to life. They also draft and revise their pages, adapting both text and code to changing designs and changing browsers. As with any writing, Web pages also benefit from getting others' reactions and then incorporating their questions and suggestions into new versions.

Yet, as we'll discuss in more detail in the opening chapters of this text, Web pages also offer many possibilities that are difficult to do in conventional writing situations, including color, graphics, and page design. This book is designed to help you realize those possibilities.

Who Should Use This Book?

Web Writing/Web Designing is intended as a supplemental text for any class in which Web pages are assigned, including composition and technical writing classes, as well as classes in other subjects. It can also be used by readers who are interested in designing Web pages on their own time, outside an organized class.

In organizing this book, I assumed that those who would use it already knew something about writing words (although perhaps not much about writing words for the computer screen); you may also know something about using images or laying out pages on paper. But I assume here that most readers will know little or nothing about writing HTML code, the basis for creating Web pages.

WRITING HTML

HyperText Markup Language (HTML) is the code in which Web pages are written. It's possible to use software to create the HTML code for you, but, as I point out in Chapter 1, it's not really necessary. Learning HTML gives you much more control over your pages, and much more understanding of how and why pages work the way that they do. For that reason, certain chapters in this book focus on learning HTML basics, and others discuss more general ideas about writing words or composing images or designing pages for a computer screen. The two processes aren't really separate—writing code is part of the process of writing pages in general. However, it may be simpler to handle the two processes separately; that is, it may be easier to understand the code first and then work on the words and pictures. Still, if you want to take a hypertextual approach and work on code, words, and pictures simultaneously, it's possible to move through the explanations of HTML fairly quickly.

Each chapter includes a writing assignment, as well as exercises designed to give you a chance to try out the codes covered in the chapter. With each assignment are a few tips about screen writing and design as opposed to paper writing and design. Two Web site assignments later in the book will also let you work on more complex designs. You can also start with these site assignments and then use the exercises and chapter assignments to work on segments of each one if you wish.

Web writing requires a certain amount of technical information in order to function: graphic formats, browser-safe colors, usable fonts, and so forth. Much of this information is included in boxes throughout the chapters (when it isn't included within the chapter text). Again, once you have the general idea of how these parts of Web pages work (for example, when to use a GIF graphic rather than a JPEG), you can move on to more interesting topics, such as choosing those graphics and making them effective.

Finally, at the end of each chapter, I include a bibliography of Web page URLs that provide supplemental information for those interested in going further with the topics covered in the chapter.

THE WEB SITE

The *Web Writing/Web Designing* Web site at http://www.abacon.com/batschelet/ provides online versions of all the code given in the text so that you can see what the illustrations look like on screen as well as in print. You can also download versions of the HTML files to work with if you wish. In addition, there are links to all of the examples included in the text, as well as the URLs provided in the bibliographies at the end of each chapter, so you can see the original pages in full color and get the benefit of expert advice.

Also included are several ways in which you can give me feedback and discuss the book with others who are using it, including a bulletin board and my e-mail address.

Finally, the Web site includes a section for teachers who are using *Web Writing/Web Designing* in class, with some suggestions for teaching Web writing.

CONTENTS

It's difficult to devise a linear sequence for an activity such as Web writing that doesn't have its own clear-cut sequence. Who's to say you won't want to write links as soon as you learn to write text, for example? So, although these chapters build upon each other to some extent, you can also use them out of sequence if you prefer, moving to linking and site organization before you tackle design, for example.

Web Writing/Web Designing is organized like this:

Chapter 1: Entering the Web: An Introduction to HTML provides an overview of Web writing and a justification for learning HTML. Although authoring programs provide Web writers with several helpful features, there are many sound reasons for learning basic HTML code even though you may write many of your pages using authoring software. This chapter explains some of those reasons. Finally, the chapter provides a brief introduction to HTML basics.

Chapter 2: On the Web in an Hour: Your First Web Page includes the conventions of a basic Web page, including text and color codes. It also includes some necessary (but sometimes overlooked) supplemental information, such as how to save files and the technicalities of Web color.

Chapter 3: GIFs and JPEGs and PNGs, Oh My: Using Images covers the basic information necessary for placing graphics on your Web pages, including locating graphics, converting them to Web formats, and placing them on Web pages. Sidebars include information on copyright, file size, resolution, naming files, and screen size.

Chapter 4: How Should I Say It: Writing for the Web discusses some guidelines for writing text on the Web, including considerations of audience and purpose, writing style, and organization. The chapter includes information and advice from a variety of authorities on Web writing, including Jakob Nielsen, Jeffrey Zeldman, and the *Sun Microsystems Guide to Web Style*.

Chapter 5: Laying It All Out: Tables explains the use of tables for design grids on Web pages. The chapter covers table construction in HTML; the use of tables for various page layouts, including creating columns and white space; as well as the use of tables to place text and graphics.

Chapter 6: How Should It Look: Web Page Design presents some principles for designing attractive and interesting Web pages, including advice from designers such as Raymond Pirouz and Kevin Mullet.

Chapter 7: Making the Web a Web: Links describes how links work in HTML, including URLs, absolute and relative links, e-mail, anchor links, and link rhetoric—ways to make your links more effective. The chapter also includes a discussion of HTML lists.

Chapter 8: How Should It Work: Site Structure explains how to organize a site, including splash pages and home pages, site architecture, and constructing navigation bars.

Chapter 9: The Future: Cascading Style Sheets provides an overview of the new developments in HTML, with instructions for using Cascading Style Sheets to format text.

The Appendix provides basic instructions for creating and using frames. The Glossary provides brief definitions of many of the terms used in this book, and the Bibliography offers sources for more information on Web page design.

WHAT? ME, CODE?

The central message stressed here and elsewhere in this book is that you can control your own Web pages, your own messages, and your own design. You don't have to rely on software to create them for you or on other people to come up with ideas. Web writing offers you resources you may not be able to find as easily anywhere else, along with a potential audience of thousands. Once you understand the basics of HTML and Web design, you can use those resources relatively easily. Then as you master those basics, you may want to do more. Chapter 9 gives you some indication of where the Web is heading next with a brief overview of Cascading Style Sheets, a new addition to HTML that is gradually being integrated into versions 4 and 5 Web browsers.

HTML may seem clumsy at first: finding missing close tags and backslashes can be exasperating, and you may find your ambitions quickly outrun your expertise. But like most skills, writing Web pages becomes easier with repetition. You'll learn why some things work the way you expect them to and why others don't by doing the same procedure two or three times until it becomes routine. Thus you shouldn't think of this project as one Web page or site; think of it as the beginning of a series of pages, a series of sites. Ideally, once you get the hang of writing a Web page, you'll want to keep going and improving, creating more and more pages with more and more complexity. My fifteen-year-old son, for example, has posted three different Web sites, each one more complex than the last. As he learned new techniques, he designed new sites on which to display them.

A number of topics are not covered here—JavaScript, forms, and databases, among others. Yet what is covered will be enough to get you going and to help you create interesting and functional Web sites. Who knows, maybe doing that will be enough to make you want to know even more. But that's another book.

ACKNOWLEDGMENTS

Web Writing/Web Designing was a fascinating but daunting book to write, and I want to thank all the many people who helped me to write it. My friend Carol Ann Britt encouraged me to come up with the original concept and then served as a test case for my instructions. Susan Romano at UTSA gave me invaluable feedback on Web teaching. The members of my Communication 4433 class were both guinea pigs and collaborators: their contributions to my chapter on Cascading Style Sheets were invaluable. I am grateful to my editor at Allyn and Bacon, Joe Opiela, for his support, and to his assistant Kristen Desmond, for helping me with the technicalities of seeing the book into print. Many thanks to the following reviewers of this edition's manuscript for their helpful comments: Glenn Blalock, Texas A&M–Corpus Christi; Anne Bliss, University of Colorado; Michael Day, Northern Illinois University; Dean Fontenot, Texas Tech; Michael Meeker, Winona State University; Irvin Peckham, University of Nebraska at Omaha; and Peter Sands, University of Wisconsin–Milwaukee. And finally, as always, I thank my family for putting up with me: Bill, Josh, and Ben, the Mac Ace.

M.B.

WEB WRITING/
WEB DESIGNING

CHAPTER 1

ENTERING THE WEB
An Introduction to HTML

WELCOME TO WEB WRITING

You have to write a Web page. Maybe you've been assigned to do it for a class, or maybe you've been asked to do it for a group you belong to. Perhaps you've even volunteered to do it for your job. Or maybe you just want to write one. Perhaps the "have to" part comes from your own determination to get out there with everybody else on the Web, to participate in the personal, political, and social life of the Internet. Whatever your motives in writing a page, this book is designed to help you do it; the text discusses the ins and outs of HTML (HyperText Markup Language), some guidelines for creating text and graphics, and some new Web developments, such as Cascading Style Sheets. But before we do any of that, we need to address some basic questions: just what is the Web and why should you want to go there?

WORLD WIDE WEB 101

Before the Web there was the Internet; actually, of course, there's still the Internet, of which the Web is a major part. The Internet is a global network of computer users; as of 1998, there were estimated to be more than 100 million users on the Internet, and that number has undoubtedly increased since then. Originally designed as a medium for collaborative research, the Internet has become an immense communication network, linking people around the world to sources of information and commerce and allowing these people to use words and images to convey their own personal, professional, and political information.

The World Wide Web is one part of the Internet, the part that supports documents formatted in HTML. These documents are read through *browser* programs that run on individual computers; the most well-known browsers are Netscape Navigator and Microsoft Internet Explorer. Several other browsers exist, some of which are text only, meaning that they won't show graphics included in Web pages.

1

WHY PUBLISH ON THE WEB?

So why should you want to go out on the Web yourself, besides just the sheer fun of it? To begin with, there are those 100 million–plus computers, all hooked up to the Web and reading Web pages. If you have information to get out or a product to sell, those 100 million users represent a great resource. Moreover, unlike newspapers, magazines, and other print media, publishing to the Web has no middlemen to get through—all you need is a computer and an Internet Service Provider so that you can mount your Web pages on an Internet server.

What does this mean to you? First of all, Web pages provide easy access to resources that are sometimes difficult to use in print, such as color and graphics. If you add a color photograph to an essay or report, in order to print it effectively, you need the following: a digital version of the photograph so that you can place it on the page (which means either using a scanner or having the photograph on a photo CD), a particular type of printer that handles color, and a certain type of paper so that the color will be reproduced at an acceptable quality. To place a color photograph on your Web page, all you need is a digital version of the photograph (to be fair, you might also need access to a graphics program, such as Adobe Photoshop, Macromedia Freehand, or Corel Draw, to convert it to the right format). Color for your page background or for your text is even easier on the Web: you just write the HTML. On paper you still need that printer and that paper to make it work effectively.

But there's an even more important reason to get into Web writing: the Web provides you with an unparalleled opportunity for communicating with your peers and others. True, you can talk to other Web users through e-mail and chat rooms, but you can do so much more through a Web page. You can share your knowledge and enthusiasm about certain topics or pass on your concern about others. You can experiment with ways of presenting information, such as using various types of graphics and layout to enhance your message, and get responses from your audience. An essay you write for a class can be seen by the members of your class (if you reproduce it for them) and perhaps by others if you or someone else distributes the copies. But your Web page is available to anyone with access to the Web; the possibilities for communication are immense. If you're careful to include your e-mail address on your pages, you can also establish a dialog with the people who visit your pages.

Publishing on the Web is relatively easy; if you have an Internet Service Provider (e.g., the company or organization where you have an e-mail account), chances are you have access to space for a Web site. After you create your pages, you send them to the Internet Service Provider, usually via a modem, and they'll mount them for you on their computer or *server*. You can also post your pages at free sites available at providers such as Yahoo!, Excite, Xoom, and GeoCities. Once your pages are mounted, they're available to everyone surfing the Web: all your classmates, your dormmates, your family, your friends, and, of course, total strangers who happen upon your site.

Of course, as you've probably already deduced, there's a little more to it than that. First of all you have to write the pages, which means creating the HTML tags (the codes that tell browsers what to do with a page) to go with them. Then you need to attract the attention of at least some of those 100 million users, many of whom will give you no more than a few seconds to demonstrate that your pages are worth their time. That means that in most cases, words won't be enough. For the World Wide Web, *writing* actually means *designing*.

Almost without exception, Web pages require some kind of graphic support; if not pictures, then at least color and classic text design features such as white space and headings. It's fair to say that the more time you spend considering the "look" of your pages, the more likely you are to get and keep the attention of the average Web surfer.

Don't think of this as a problem, however. Remember those added resources that the Web provides. You can use all the color and graphics your heart (and your design sense) desires, as long as you prepare the graphics correctly and write the HTML tags to support them.

WHY HTML?

Right now you may be wondering why HTML keeps coming up. All you have to do is get an authoring program such as PageMill or Front Page or Dreamweaver and learn it. No messing with HTML, no fiddling with tags, no formatting graphics. Right? Well...no. Technically, these authoring programs are graphic-interface HTML editors; that is, they create a facsimile of a browser window and then the program writes the HTML for you while you place the parts of your Web page in the pretend browser. So the page still runs on HTML, even though you may not be aware of it. Authoring programs can be a big time-saver. In fact, many professional Web designers use them to create large-scale sites. But there are still good reasons to learn HTML, even if you plan on using an authoring program to create many of your Web sites.

First, knowing how HTML works will help you understand what it can and can't do, such as why you can't make your centered graphic stay where you want it to above your text (because HTML doesn't offer center alignment as one of the options with graphics).

Second, learning HTML will make it simpler for you to learn and use various authoring programs on different computers. HTML is HTML; no matter what computer you're working on—Mac or IBM or Unix—the tags will be the same. Because you write HTML in text-only files (see the Some Basic Terminology box), you can use any computer to work on them—you don't need to use the same program every time, just a program that creates text-only files and that includes text programs that came with your operating system, such as SimpleText (Mac) and Notepad (Windows). So if you created your pages in Front Page and you move to a computer that has only Netscape

Composer available, you'll still be able to function. The code will be the same, no matter what authoring program you use, and you can pick up the format of another authoring program more easily if you understand what's going on.

Finally, with an authoring program, you're limited to what the program can do: maybe it's based on HTML 3.2, which has fewer capabilities than HTML 4.0. Perhaps the program doesn't have any means to insert Cascading Style Sheets because that's a different version of HTML. Maybe the program uses an awkward combination of tags to create a particular effect because that's what was being done when that version of the authoring program was written (such as using repeated BLOCKQUOTE tags to indent text), even though the current version of HTML (and current browsers) has simpler, more elegant alternatives. In other words, if you know only an authoring program, you can do only what that authoring program can do. If you know HTML, you can do what *you* can do (provided that HTML will let you do it!).

Authoring programs are very useful, and you may want to learn or own one. Once you know HTML, you can use the authoring program to do time-consuming and tag-heavy tasks such as constructing tables or frames. Authoring programs can also do some things more quickly than you can, such as writing the tags for a complex page line by line, and if you're constructing a multipage site, that can be a big time-saver. In addition, most authoring programs allow you to edit the HTML as well as use the pretend browser windows to construct pages. (If you're going to purchase an authoring setup, make sure it will allow this.) Being able to edit the HTML will allow you to change parts of the markup and add any tags that the authoring program didn't. Learning HTML and then using an authoring program for speed and simplicity is a very effective way to go about Web page construction. After you finish this book, you should be able to work with most of the authoring programs available either commercially or for free.

But you don't need an authoring program to get started. Here's the software you do need. (We'll take it as a given that you have access to a computer.)

- A text program (i.e., a word processing program or a writing program such as SimpleText or Notepad) that will write text-only files (see the Some Basic Terminology box)
- A graphics program that will create and/or edit graphics in GIF and JPEG formats (and maybe PNG)
- A Web browser such as Netscape or Microsoft Internet Explorer

In fact, although graphics are essential in creating effective Web pages, you can actually start your Web-writing career without that graphics program. But the text program and the browser are mandatory, as you'll see.

WHAT IS HTML?

So what is HTML, and why is it necessary? One of the ideas behind the World Wide Web is that all computers connected to the Web should be able to read any Web page posted on any server. In order for that to happen, all these pages have to be written in a common language—not necessarily the language spoken by the owners of those computers, but a common formatting language: HTML.

■ SOME BASIC TERMINOLOGY ■

Tag. A command that specifies how part of a Web page should look or behave. For example, `<B>` is a tag that specifies that a word or words should be shown in boldface.

Text-only file. ASCII code without any of the formatting provided by a word processor. Text-only files are readable on most computers and don't rely on word processing software for translation.

Mount (upload). To transmit Web pages to an Internet server.

Server. A computer that manages network resources, providing access to files, databases, and so forth. A Web server supports documents formatted in HTML and provides access to them for users of the World Wide Web.

HTML is a "tagging" language. It's a subsection of a much larger and more complex language, SGML (Standard Generalized Markup Language), used to make complex technical documents readable in a variety of formats. The idea behind HTML is that the elements of your page—text, graphics, and other components such as video clips or sound files—will all be formatted using a limited number of tags to identify content and style. If everyone uses the same tags, then all browsers ought to be able to read the pages that use those tags, no matter what type of computer they were created with or what type of server they're mounted on.

To a certain extent, that's how HTML has worked out; the basic HTML tags are readable by all browsers, although they may not be reproduced in exactly the same way. For example, if you place your text inside these tags—`<EM></EM>`—in most browsers, it will be italicized, but in a few browsers (ones that don't recognize graphics, for example), it will be underlined. If you place your text inside these tags—`<I></I>`—it will be italicized in all the browsers that recognize the tags but will be reproduced as plain text in the browsers that don't.

However, some browsers (Netscape and Explorer, for two) have introduced tags that can be read only by those particular browsers. Because using tags that can be recognized only by one particular browser defeats the whole purpose of a universal tagging language, these tags are not included in this text. However, some authoring programs do include them. You can usually find out whether these tags are readable in all browsers by consulting a Web reference (either printed or online—check the URLs and references at the end of this chapter); just be aware that some browser-specific tags do exist.

Finally, the most recent innovation in Web design, Cascading Style Sheets (CSS), has yet to be fully implemented in any of the browsers currently in widest use (at this writing, Netscape 4.x and Explorer 5.x). However, CSS has been adopted by the World Wide Web Consortium, the Web governing body, and it will be fully implemented eventually. (As this book was being written, Netscape announced that Netscape 6.0 will fully implement CSS, and Microsoft has said the same about IE 5.0 for Macintosh.)

But for now, let's look at some basic information about writing HTML.

HTML BASICS

All HTML tags are written between less than and greater than signs (angle brackets) that look like this:

    <B>

Container Tags

Many HTML tags are *container* tags; that is, they have both an opening and a closing tag.

<B> Bolded Heading </B>

Opening Tag   Closing Tag

(In a browser it would look like this: **Bolded Heading.**) The text (Bolded Heading) is *contained* between the two tags; all the text that appears between these tags will be affected by them. Think of the opening tag as a switch that turns on the effect and the closing tag as a switch that turns off the effect.

The opening and closing tags use exactly the same word or symbol (e.g., B), but the closing tag has a slash before the word or symbol (e.g.,/B) in order to differentiate it from the opening tag. If you're using one of these container tags, *you must include the closing tag as well as the opening tag.* If you forget and leave off the close, your text may not show up in the browser, or it may not have the style you expect it to have because you didn't turn the effect off after you turned it on. Think of it as a refrigerator door—if you forget to close it, the contents may not live up to your expectations!

Not all HTML tags are container tags; some (such as the line break tag, <BR>) require only an opening tag. (It's still called an *opening tag,* even if there's no closing tag, because it opens the HTML command.) If you don't remember whether the tag you're using needs a closing tag, check the tag tables provided at the end of Chapters 2, 3, 5, and 7.

Attributes

Some tags also include *attributes.* These words or symbols affect the appearance or behavior of the elements on the page. Here's an example.

<FONT COLOR="red">
Tag   Attribute   Value

Here <FONT> is the opening tag (there's also a closing tag: </FONT> that comes at the end of the section where you want to change the font color) and COLOR is the attribute that indicates what aspect of the font you want to deal with. More than one attribute can be included in a tag; for example, you could have

<FONT COLOR="red" SIZE="4">

However, all the attributes must come after the opening *tag* word (in this case FONT) and before the closing angle bracket. Not all tags can include attributes; we'll cover the ones that can as we go through the HTML tags.

Values

Attributes frequently have *values;* that is, they include further information such as color, size, and location. The value part of the attribute comes after the equal sign, such as this:

<FONT COLOR="red">
Tag   Attribute   Value

Values are frequently, although not always, placed in quotation marks; however, because including quotation marks won't make any difference to the values that *don't* require them, whereas leaving the quotation marks off will make a big difference to the values that *do* require them, enclose all values in quotations marks just to be on the safe side.

▓   QUOTATION MARKS IN HTML   ▓

The quotation marks you use in HTML must be *straight* quotation marks rather than the curly *"smart"* quotation marks (like the ones I just used around *smart*). If you have the smart quotation marks option checked in your word processor (and you plan on using your word processor to write HTML), uncheck it while you work on your Web page.

■ TRY IT

Try writing some of these tags now, just to get the hang of it. All Web pages include the following tags at the beginning and end. (We'll talk more about this in the next chapter.) Write them out for yourself now; then save them as a text-only file called HTML Template.

```
<HTML>
<HEAD>
<TITLE></TITLE>
</HEAD>
<BODY>
</BODY>
</HTML>
```

Nesting Tags

Frequently, you'll have a series of tags around a particular paragraph or text section. For example, you might have

```
<B><I>Attention Must be Paid</I></B>
```

(In a browser it would look like this: ***Attention Must be Paid***.) The main thing to remember here is that *the tag that opens first closes last.* In other words, if B is the first opening tag, then it should be the last closing tag as well. Your closing tags should repeat your opening tags in reverse order. Not all tags can "nest" like this, but several can.

WRITING YOUR HTML

As indicated earlier, you can write your HTML in any kind of text program, from a full-featured word processor to the simple "text-only" programs that come with your computer's operating system (Notepad for Windows; Simple-Text for Macintosh). There is one big advantage to using these simple text-only programs, however: *your HTML must be saved as a text-only (ASCII text) file.* If you forget and save your HTML as a Microsoft Word file or a WordPerfect file or any file other than text only, the browser will not be able to read your code.

Because Notepad and SimpleText are both text-only programs, you can't slip up and forget to save the file as text only. Some HTML editing programs such as BBEdit or HomeSite also produce text-only files; if you're not sure in what format your files are being saved, check the documentation for the program.

Spacing HTML

Your browser won't recognize spaces between your tags. Code written like this:

```
<H1>HTML and Me</H1>I didn't start out writing
HTML. I began writing Web pages a couple of years
ago, using an authoring program because I thought
HTML was difficult to learn and use. <P>I was
frustrated, however, because the authoring
program didn't do what I wanted it to do. After
reading several books on Web design, I decided to
try writing my first HTML.</P><P>Surprise: HTML
isn't that hard—and it's fun!</P>
```

will look exactly the same in the browser as code written like this:

```
<H1>HTML and Me</H1>

I didn't start out writing HTML. I began writing
Web pages a couple of years ago, using an
authoring program because I thought HTML was
difficult to learn and use.

<P>I was frustrated, however, because the
authoring program didn't do what I wanted it to
do. After reading several books on Web design, I
decided to try writing my first HTML.</P>

<P>Surprise: HTML isn't that hard—and it's
fun!</P>
```

The big difference between these two examples, however, is that the second one is a lot easier to review quickly. For example, suppose I left off the / on the H1 close tag. If I tried to look at the text in a browser, it would look like this:

HTML and Me

I didn't start out writing HTML. I began writing Web pages a couple of years ago, using an authoring program because I thought HTML was difficult to learn and use.

I was frustrated, however, because the authoring program didn't do what I wanted it to do. After reading several books on Web design, I decided to try writing my first HTML.

Surprise: HTML isn't that hard—and it's fun!

Everything on the page would come up in bold at a large size (the H1 heading style) because I forgot to turn off the H1 tag. I'd want to "debug" that code to find out what went wrong. It would be a lot harder to find the error in the first example, where all the lines run together, than in the second example, where each paragraph is set off.

Remember that the browser won't notice vertical spaces you insert in your text, either. If you want to put space between lines or between paragraphs on your Web page, you'll need to insert a tag that will create that space (usually either <P> or <BR>) rather than just hitting the return key as you would on a word processor.

HTML Tags

HTML isn't case sensitive: browsers will recognize your tags whether they're in capital letters or in lowercase. However, many HTML writers put their tags in all caps anyway for the same reason that they put some vertical space between their paragraphs and command sections: it's easier to debug that way.

If you put your HTML tags in capital letters, they'll be set off from your text so that you can find the tags quickly to check them for errors. However, if you don't like hitting the shift or caps lock key each time you write a tag (and taking it off to write the angle brackets), you can write them lowercase. Whatever works for you!

The best way to learn HTML is to start writing it, which we'll do in the next chapter. Keep this in mind: HTML requires trial and error. Your first attempts may seem less than stellar. However, once you get the hang of it, it won't seem as difficult. Take a problem-solving approach: how can you get this resource to do what you want, and how can you use this resource to make your message more exciting? HTML offers a wide range of possibilities, so let's start exploring!

At the end of each chapter is a listing of some Web sites for further information regarding the chapter topics. You can enter the URLs in your Web browser, or you can do it the easy way and go to the Web site that's associated with this text at http://www.abacon.com/batschelet/, where you'll find links to the pages cited here.

■ TRY IT

One of the best ways to learn about HTML is to see how other Web writers use it. Find a page that you like on the Web, one whose design you're particularly fond of. Then take a look at the HTML code. In Internet Explorer, go to the View menu and select Source. In Netscape, go to the View menu and select Page Source. What comes up on the screen will be the HTML code that makes the page work. As you go through the following chapters and begin writing your own Web pages, you can do this frequently. Go to others' pages and see how those pages were created. Then see if you can use similar techniques on pages of your own.

WEB SITES

http://bignosebird.com/newbie.shtml
Big Nose Bird's page of reassurance for absolute beginners

http://info.med.yal.edu/caim/manual/index.html
Yale Center for Advanced Interactive Media Style Guide: probably the most austere and elegant style guide on the Web

http://www.boutell.com/faq/oldfaq/index.html
Tom Boutell's FAQs (Frequently Asked Questions) about the Web and Web authoring

http://www.december.com/html/
HTML Station, a site for all things related to HTML

http://www.ncsa.uiuc.edu/General/Internet/WWW/
      HTMLPrimerP1.html#GS
An introduction to HTML from the National Center for Supercomputing Applications

http://www.sun.com/styleguide/
Sun Microsystems Guide to Web Style

http://www.w3.org/History.html
A short history of the World Wide Web from the World Wide Web Consortium, the Web's governing body

http://www.w3.org/MarkUp/MarkUp.html
World Wide Web Consortium's HTML home page

CHAPTER 2

ON THE WEB IN AN HOUR
Your First Web Page

BASIC HTML TAGS

Basic HTML tags are easy to write; in fact, you can learn enough HTML in an hour to have a basic Web page ready to post. It won't necessarily be anything to write home about, but it will be mountable! This chapter covers how to begin and end a simple page, how to set up some simple colors, and how to enter text (along with some basic formatting).

SETTING UP YOUR PAGE

Every HTML page includes the same basic informational skeleton; in fact, you can even set up this information as a template (a pattern page that you can use and reuse) because you'll use it repeatedly. If you did the exercise in Chapter 1, you already have this template saved.

Beginning Your Page

1. At the top of the page, type <HTML></HTML>
2. After <HTML>, but before </HTML>, type <HEAD></HEAD>
3. After <HEAD>, but before </HEAD>, type <TITLE></TITLE>
4. After <TITLE>, but before </TITLE>, write a brief, descriptive page title. (See the following information for more about titles.)
5. After </HEAD>, but before </HTML>, type <BODY></BODY>

Your page so far should look like this:

```
<HTML>
<HEAD>
        <TITLE>My First Page</TITLE>
</HEAD>
<BODY>
```

```
</BODY>
</HTML>
```

Now let's look at these tags individually.

< HTML > < /HTML >

All HTML pages begin and end with these tags to tell the browser in what language the page is written. In fact, many browsers will read the page without these tags, but it's a good idea always to include them anyway. Given the number of browsers (and the number of browser versions) on the Web, it never hurts to be specific about your pages.

< HEAD > < /HEAD >

All HTML pages are divided into head and body sections. The head section doesn't appear in the browser window, but it includes information the browser needs in order to identify the page, such as the page's title; that is, the text that appears at the page's top (see Figure 2.1). It can also include other information, some of which we'll discuss later in this book.

< TITLE > < /TITLE >

The title of the page appears in the title bar above the browser window. Don't confuse the title with a heading—headings require a different format. In the

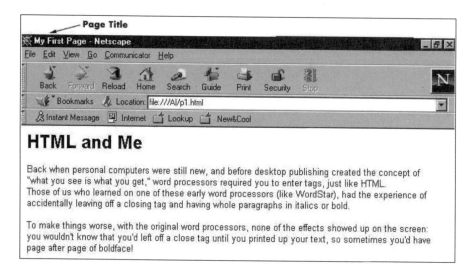

FIGURE 2.1 Page Title

example in Figure 2.1, My First Page is the page title; in the browser, it shows up as My First Page—Netscape (in Macintosh it will be Netscape: My First Page).

Titles are used by many search engines on the Web to identify the subject and keywords associated with your Web site pages. Because you want users to be able to find your pages (which means you want search engines to identify the subject of your pages correctly), you need to write your titles carefully.

Some things to keep in mind about titles:

- Keep your titles short, but make them specific. (For example, Page One doesn't tell your reader much; a word or two that identify your subject, such as Soccer Rules, will be more helpful.)
- Don't include any formatting (e.g., italics or bolding), links, or images in your titles; each browser will supply its own formatting.
- Don't use colons or backslashes; these symbols are used by some operating systems for other purposes.
- If you have more than one page, you can tie them together by using a common element in each title. (For example, if you were doing a Web site for a soccer team, you could begin each page title with Soccer, such as Soccer Rules, Soccer Tips, etc.)

< BODY > < /BODY >

The body section of your page includes everything that visitors to your Web site will see in the main body of the page below the title bar: the text, images, and links to other pages. The bulk of your work will be done on the body section.

A tag table of the tags covered in Chapters 2, 3, 5, and 7 is located at the end of each chapter. If you lose track of what something means, check these listings for a quick explanation. (Also remember the Web page URLs listed at the end of each chapter that provide more help and advice.)

FILLING IN THE BLANKS

Now that you've got the basic page skeleton, you can go on to fill in the rest of the page.

Paragraphs and Line Breaks

Paragraphs in basic HTML are not indented (although we'll see a way to indent them using Cascading Style Sheets in Chapter 9); they're separated by a line of vertical space. This paragraph format looks much like the conventional business letter format. The default text is single-spaced.

Creating a Paragraph

1. After <BODY> but before </BODY>, type <P>
2. Type the text of your paragraph.
3. Type </P>

Technically the paragraph tag, <P>, isn't a container tag; that is, browsers will read <P> without the </P>. However, when you use Cascading Style Sheets to define styles for your paragraphs, you need to use </P>. Thus it's a good idea to get into the habit of using the <P></P> whenever you want to create a paragraph.

Sometimes you may want to create a new line of text without having the line of vertical space you get when you use the paragraph tag. For example, if you were writing a poem in HTML, you might not want extra space between each line. In that case, you can use the line break tag <BR>; this tag will start a new line without adding any extra space between the lines.
 is not a container tag, so you don't need to use </BR>.

Creating a Line Break

1. At the end of the sentence (or text line) where you want a break, type
2. Type the next line of text.

Sometimes novice Web writers try to create extra space on a page by using multiple line breaks such as this:

. Unfortunately, this doesn't work: the browser will simply collapse all those line breaks into a single line break. We'll discuss some ways to create white space later on when we discuss adding images in Chapter 3.

■ TRY IT

Write the HTML and text for a Web page that includes the following:

1. A paragraph <P>
2. At least three lines of text that end in line breaks

Remember to include the basic template: <HTML></HTML>, < HEAD></HEAD>, <TITLE></TITLE>, <BODY></BODY>

You can start working on the Web page for the first Web page assignment on page 68 if you wish.

Headings

In HTML there are six levels of headings, from <H1> (the biggest) to <H6> (the smallest).

Creating a Heading
1. After <BODY> but before </BODY>, type <H1>
2. Type a brief heading.
3. Type </H1>

If you want to create a smaller-sized heading to use instead of <H1>, type <H2> or <H3>, and so forth, and instead of </H1>, type </H2> or </H3>, and so on.

All the headings will be placed against the left margin, unless you indicate otherwise by using an ALIGN attribute. Use the following instructions if you want to change the alignment of the heading.

Realigning a Heading
1. After <BODY> but before </BODY>, type <H1>
2. Type ALIGN= and either LEFT, RIGHT, or CENTER
3. Type >
4. Type a brief heading.
5. Type </H1>

These headings are preformatted by the HTML code: they'll be bolded and sized according to the type of computer you're using to view the page. (Type looks slightly smaller on a Macintosh than on a PC.)

Headings have an automatic line break afterward, so you don't need to end the line with <BR> or begin the next paragraph with <P>. Just start typing your text after </H1>.

■ TRY IT

Write the HTML and text for a Web page that includes the following:

1. Two headings in different sizes
2. Two paragraphs

Remember to include the basic template: <HTML></HTML>, <HEAD></HEAD>, <TITLE></TITLE>, <BODY></BODY>

You can add the text you write here to the text you wrote for the previous exercise and continue working on the Web page assignment if you wish.

Depending on how much text you've entered, your page should look something like this:

```
<HTML>
<HEAD>
        <TITLE>My First Page</TITLE>
</HEAD>
<BODY>
<H1>HTML and Me</H1>
Back when personal computers were still new, and
before desktop publishing created the concept of
"what you see is what you get," word processors
required you to enter tags, just like HTML.<BR>
Those of us who learned on one of these early word
processors (like WordStar), had the experience of
accidentally leaving off a closing tag and having
whole paragraphs in italics or bold.

<P>To make things worse, with the original word
processors, none of the effects showed up on the
screen: you wouldn't know that you'd left off a
close tag until you printed up your text, so
sometimes you'd have page after page of
boldface!</P>
</BODY>
</HTML>
```

In a browser it looks like Figure 2.2.

If you realign the heading using `<H1 ALIGN="center">HTML and Me</H1>`, the result in a browser would look like Figure 2.3.

CHECKING WITH A BROWSER

As you look at Figure 2.2 and Figure 2.3, you'll see that they look slightly different. That's because two different browsers on two different computers were used to view them. That's one of the reasons you should check your pages frequently to see how they're showing up in the standard browser windows. Take this opportunity to look at the pages you wrote for the two previous exercises, after you've finished typing out all of your text and tags. You're not going

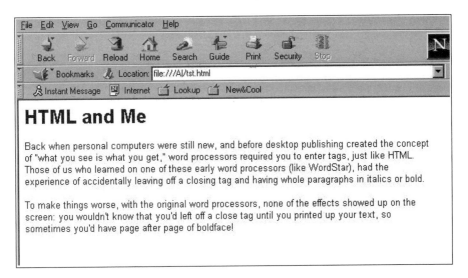

FIGURE 2.2   First Page

online because your page isn't mounted yet; instead, you'll look at the page off-line, which will give you a fairly clear idea of how it would look on the Web.

Looking at Your Page in a Browser

1. Save the file, using the extension `.html` (unless you're using Windows 3.1 or an earlier version of Windows—then it's `.htm`).
2. Open your Web browser.
3. In Macintosh: in Netscape, go to File/Open/Page in Navigator and locate your page; in Explorer, go to File/Open File and locate your page.

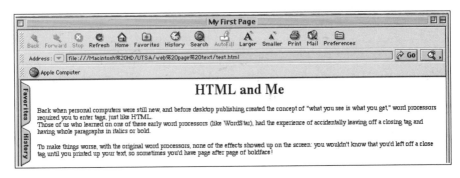

FIGURE 2.3   Centered Heading

In Windows: in Netscape, go to File/Open, press the Choose File button, and locate your page; in Explorer, go to File/Open, press the Browse button, and locate your page.

4. Open your page on your screen.

Looks pretty bland, doesn't it? Let's see what we can do to spice it up a little.

■ NAMING AND SAVING FILES ■

You must save your HTML file before you can look at it in a browser, but you need to be aware of some conventions to naming. First of all, to repeat: your HTML files need to be saved in text-only (ASCII text) format. If you save them in Microsoft Word format or WordPerfect format or *any* format other than text only, they won't be readable on the Web. You must also use the `.html` or `.htm` extension on the name of your HTML files. Extensions are a series of letters at the end of a file name; usually they're separated from the file name by a period— e.g., `soccer1.html`. The extension tells the browser what kind of information is stored in the file, so `.html` indicates that the information is written in HTML, whereas `.gif` indicates that the information is a GIF graphic. (For more on graphics, see Chapter 3.) Although it's sometimes assumed that PC users save as `.htm` and Mac users save as `.html`, the distinction is actually the version of Windows that PC users are using (or not using). If you're using Windows 3.1 or earlier, you can't use a four-letter extension such as `.html`—you must use `.htm`. But if you're using a later version of Windows or a Mac, you can use `.html`.

When you name your files (including HTML, graphics, sound, etc.—any files), there are two things to remember:

1. Keep the file names short: usually no more than eight characters. Some browsers can't read longer file names.

2. Leave *no spaces* between words in your file names. A name such as `Soccer 1.html` wouldn't work as a file name. You'd have to make it `Soccer1.html`. Again, you need to be careful about this because some browsers won't read files that have spaces in the names.

ADDING COLOR

It's easy to add color to both your page text and your page background, although there are some complex issues about color that are addressed later. (Remember, you're just trying to get up and running in an hour here!) Let's begin with the background.

▤ COLOR ON THE WEB ▤

Throughout the text you'll be advised to use either *hexadecimal colors* or sixteen predefined color names: **silver, gray, maroon, green, navy, purple, olive, teal, white, black, red, lime, blue, magenta, yellow,** and **cyan.** Web color is a confusing issue, but the bottom line is this: Web browsers use only about 255 colors. If you use a color that isn't included in those 255, the browser will *dither* your image; this means that the browser will substitute one of the 255 colors that may or may not show up the way you intended. The solution is to use one of the colors recognized by the browser: a hexadecimal color.

Hexadecimal color numbers always look like this: #FFFF00 (it's a nice, bright yellow in reality)—they always begin with # and have six letters and/or numbers. The numbers convert RGB colors (red, green, and blue—the standard color screen for computer monitors) to mathematical equivalents that Web browsers can understand. Although several books and Web sites will explain what each pair of numbers refers to (see the URLs at the end of the chapter), it's easier simply to look at a hexadecimal color chart and enter the number for the color you want. Several sites on the Web will allow you to download a hexadecimal chart; many graphic and Web-authoring programs also include them. However, remember when you're creating graphics that you need to use the right colors: otherwise the appearance of your page may not be what you expected.

BGCOLOR (Background Color)

HTML has sixteen predefined colors, along with about 200 hexadecimal colors (see the Color on the Web box). The sixteen are **silver, gray, maroon, green, navy, purple, olive, teal, white, black, red, lime, blue, magenta, yellow,** and **cyan.** Let's use these to start with. Background color is defined as an attribute in the BODY tag.

Defining Background Color
1. After <BODY, type BGCOLOR=
2. Type the color you want for your background in quotation marks (e.g., BGCOLOR="cyan").
3. Type >

Text Color

Frequently, you need to change your text color as well as your background color. In fact, some Web designers advise that whenever you change any of the default colors on your page, you need to change them all because some users will have changed their default colors and their defaults may not show up on your new background color. (Imagine what would happen if a user had set her default text color to red and you changed the background color to red without changing the default text color.) Again, text color is one of the attributes in the BODY tag.

Defining Text Color

1. After <BODY, type TEXT=
2. Type the color you want for your text in quotation marks (e.g., TEXT="red").
3. Type >

The order in which you put BGCOLOR and TEXT doesn't matter—that is, TEXT could come first. The <BODY tag for my page now looks like this:

```
<BODY BGCOLOR="cyan" TEXT="red">
```

These attributes will affect your entire page: the background and all the text. However, you can override the text color if you want to change it for a particular section. (More information about how to do this will come later in this chapter.)

■ TRY IT

Write the HTML and text for a page that includes the following:

1. A background color
2. A text color
3. A paragraph
4. A heading

Remember to include the basic template: <HTML></HTML>, <HEAD></HEAD>, <TITLE></TITLE>, <BODY></BODY>

FORMATTING TEXT

Logical and Physical Formatting

HTML has two ways to change the style of text: (1) "logical" tags that indicate that a text should be changed but that allow the browser to dictate what that change should look like, and (2) "physical" tags that dictate precisely what the text will look like, no matter what the browser defaults are. For example

`<STRONG>` is a logical tag; in Netscape and Explorer, it's shown as boldface, but in other browsers such as Mosaic it may be shown in some other style, perhaps underlining. In contrast `<B>` will always be shown as boldface as long as the browser recognizes the tag; however, if a browser doesn't recognize the tag, the text will be shown without any formatting at all. On one hand, physical formatting allows you to make sure your text looks the way you want it to look. On the other hand, not all browsers recognize the physical tags; if you use logical tags, the text will always have some formatting, although it may not look precisely the way you intended it to.

Formatting Text Logically

1. Type `<EM>`
2. Type text you want to italicize.
3. Type `</EM>`
4. Type `<STRONG>`
5. Type text you want boldfaced.
6. Type `</STRONG>`

Formatting Text Physically

1. Type `<I>`
2. Type text you want to italicize.
3. Type `</I>`
4. Type `<B>`
5. Type text you want boldfaced.
6. Type `</B>`

Block Quotes

Using `BLOCKQUOTE` will set off a section of text, much as you would indent a long quotation in a paper. Most browsers center block quotes; however, some browsers italicize them. (Welcome to the wonderful world of the browser wars!) This is one reason not to rely on block quotes for complicated layout purposes (we'll discuss layout in Chapter 5), but you can use them for relatively simple layout changes.

Using a Block Quote

1. Type `<BLOCKQUOTE>`
2. Type any text formatting you want to add. (If you want a vertical space, type `<P>`)

3. Type your text.

4. Type the closing tags for any text formatting you're using.

5. Type </BLOCKQUOTE>

Using these formatting tags, my HTML looks like this:

```
<HTML>
<HEAD>
       <TITLE>My First Page</TITLE>
</HEAD>
<BODY BGCOLOR="cyan" TEXT="red">
<H1>HTML and Me</H1>
Back when personal computers were still new, and
before desktop publishing created the concept of
"what you see is what you get," word processors
required you to enter tags, just like HTML.<BR>
<BLOCKQUOTE>
Those of us who learned on one of these early
word processors (like <EM>WordStar</EM>), had
the experience of accidentally leaving off a
closing tag and having whole paragraphs in
<I>italics</I> or <B>bold</B>.</BLOCKQUOTE>

<P>To make things worse, with the original word
processors, none of the effects showed up on the
screen: you wouldn't know that you'd left off a
close tag until you printed up your text, so
sometimes you'd have page after page of
<STRONG>boldface!</STRONG></P>
</BODY>
</HTML>
```

In a browser, it would look like Figure 2.4. (To see the full-color version, look on the Web page associated with this text at http://www.abacon.com/batschelet/)

■ TRY IT

Write the HTML and text for a page that includes these tags:

1. <EM></EM>

2. <STRONG></STRONG>

3. <I></I>

4. <B></B>

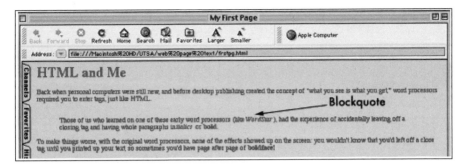

FIGURE 2.4   Formatted Text

5. <BLOCKQUOTE></BLOCKQUOTE>

6. <P></P>

Remember to include the basic template: <HTML></HTML>, <HEAD></HEAD>, <TITLE></TITLE>, <BODY></BODY>

You can add some of these tags to the text you've already produced in the previous exercises, as well as add more text if it's appropriate.

Changing Type Size and Color

As you may have noticed by now, the default settings in browsers don't always produce text at the most readable size. You can change the size of the text (i.e., the font size) by adding a tag after the **BODY** tag. The easiest way to change the font size is to change the size number. The default size is 3: to make the text larger, use a number from 4 to 7; to make the text smaller, use 1 or 2. (But be careful: 1 or 2 may not be readable on some monitors.)

Changing the Font Size

1. Put your cursor after <BODY>

2. Type <BASEFONT SIZE=

3. Type a number between 1 and 7 within quotation marks (e.g., <BASEFONT SIZE="5">).

4. Type >

There are some things to keep in mind when you change your font size. First, the **BASEFONT** tag affects only the text on the page—it has no effect on your headings. Be careful not to enlarge your text above the size of your headings, or the headings will no longer stand out.

You can use only one **BASEFONT** tag on a page. However, you can change individual sections of a page, or even individual letters, by using the **FONT** tag.

Changing the Font Size of a Section

1. Put your cursor at the beginning of the section where you want to change the size.
2. Type `<FONT SIZE=`
3. Type a number between 1 and 7, or type +n (a number between 1 and 7) or −n to add size to or subtract size from the **BASEFONT** size, in quotation marks (e.g., `<FONT SIZE="+2">`).
4. Type `>`
5. Type your text.
6. At the end of the section, type `</FONT>`

You can also use the **FONT** tag to change the type color in a particular section or a particular word or even a group of letters. The **FONT COLOR** tag and attribute are used to change the color in a section of text; the **TEXT** attribute in the **BODY** tag is used to set the color for all the text on the page.

Changing the Font Color of a Section

1. Put your cursor at the beginning of the section where you want to change the color.
2. Type `<FONT COLOR=`
3. Type the name of the color (or the hexadecimal number) in quotation marks (e.g., `<FONT COLOR="blue">`).
4. Type `>`
5. Type your text.
6. At the end of the section, type `</FONT>`

Now, with all these tags, my HTML looks like this:

```
<HTML>
<HEAD>
        <TITLE>My First Page</TITLE>
</HEAD>
<BODY BGCOLOR="cyan" TEXT="red">
<BASEFONT SIZE="4">
<H1>HTML and Me</H1>
Back when personal computers were still new, and
before desktop publishing created the concept of
"what you see is what you get," word processors
required you to enter tags, just like HTML.<BR>
```

```
<BLOCKQUOTE>
<FONT SIZE="+2" COLOR="blue">
Those of us who learned on one of these early
word processors (like <EM>WordStar</EM>), had
the experience of accidentally leaving off a
closing tag and having whole paragraphs in
<I>italics</I> or <B>bold</B>.</FONT>
</BLOCKQUOTE>

<P>To make things worse, with the original word
processors, none of the effects showed up on the
screen: you wouldn't know that you'd left off a
close tag until you printed up your text, so
sometimes you'd have page after page of
<STRONG>boldface!</STRONG></P>
</BODY>
</HTML>
```

Figure 2.5 shows what it looks like in the browser. (To get the full effect, look at the page on the text Web site at http://www.abacon.com/batschelet/.)

You can also use the `<FONT></FONT>` container to change a few letters at a time, even one letter to create a drop cap (an enlarged letter at the beginning of a paragraph).

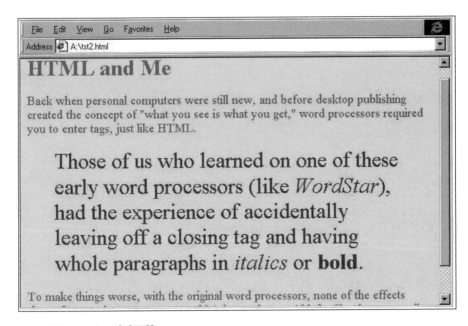

FIGURE 2.5   Special Effects

Creating Drop Caps

1. Put your cursor before the letter that will be your drop cap.
2. Type `<FONT SIZE=`
3. Type a number between 1 and 7, or type +n (a number between 1 and 7) to add size to the `BASEFONT` size, in quotation marks (e.g., `<FONT SIZE="+2"`).
4. Type `>`
5. Type the letter.
6. Type `</FONT>`

You can also change the color of your drop cap; just add `COLOR=` after `SIZE=` and put the color name or number in quotation marks. The tags for a drop cap look like this:

```
<FONT SIZE="+2" COLOR="green">B</FONT>ack
```

In a browser, it looks like Figure 2.6.

■ TRY IT

Write the HTML and text for a page that includes these tags and attributes, along with appropriate values:

1. `<BASEFONT>`
2. `TEXT`
3. `<FONT>`

Remember to include the basic template: `<HTML></HTML>`, `<HEAD></HEAD>`, `<TITLE></TITLE>`, `<BODY></BODY>`

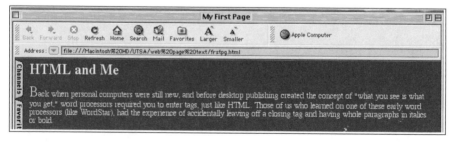

FIGURE 2.6   Drop Cap

You can add these tags to the text you've already produced for your Web page if you wish, or you can write new text if it's appropriate.

OTHER TAGS

As you can see, there's a lot you can do just with text; you can change the size, the color, the style, and to some extent the position of the text on the page. There are also other tags for formatting text beyond those mentioned here: tags for changing the font, for underlining and lining through, for making text blink. Of course, certain text-formatting tags are specific to particular browsers. These tags are not included for a variety of reasons. Some, such as the tags for changing fonts, are done more efficiently in Cascading Style Sheets. Some, such as the tags for blinking text, are universally hated. If you'd like to experiment (carefully) with some of these tags, consult the Web sites at the end of the chapter.

You now know enough HTML to create a simple Web page that could communicate some basic information. Assuming that your computer and your Internet connection are both working (a big assumption, let's face it), you probably did it in about an hour. Many Web page authors don't go much further than this; it suits their purposes. If you haven't looked at your HTML page in a browser, do it now.

If you're a Web surfer yourself, you may not be too impressed by what you see. That's because few effective pages on the Web are text only. (If you're not a Web surfer, you should become one, so that you can get ideas for your own pages.) Like it or not, to make a good Web page, you need to use images and design along with your text, and that's what we'll cover next.

■ WRITING ASSIGNMENT

The assignment for your first page is very simple: you'll write *three* paragraphs (`<P></P>`) about a subject of your choosing (your instructor may choose to focus your topic further). One appropriate topic might be your first experience with a computer, or your first time on the Web. Or you could write about why you want to write for the Web—what kinds of pages and topics you'd like to produce and share. In general, however, keep it simple. It's difficult enough getting comfortable with HTML without also having to be innovative with your writing at this point. Besides the two paragraphs, this first page should also have the following:

- All the correct template tags in the correct order: `<HTML></HTML>`, `<HEAD></HEAD>`, `<TITLE></TITLE>`, `<BODY></BODY>`

- At least one heading <Hn></Hn>
- A colored background
- Colored text
- At least one change in text size (other than the heading)
- Use of boldface or italic text (besides the heading)

Tips: Although text on the Web is written in a variety of styles (reflecting the variety of authors' purposes), you'll find a great deal of first-person writing there. Indeed, as you'll see in Chapter 4, some Web writers argue that first person has a distinct advantage in this context. It personalizes the relationship between Web writer and Web user, and it allows you to use a more relaxed, conversational style. Try using first person on this first page—talk about your reactions to this new process. If you hear your own voice issuing from that wilderness of tags, you may feel a little more at home!

TAGS USED IN CHAPTER 2

OPENING TAG	CLOSING TAG	EFFECT
<HTML>	</HTML>	Begins and ends page
<HEAD>	</HEAD>	Opens and closes head section of page
<TITLE>	</TITLE>	Designates title that appears in title bar of the browser window
<BODY>	</BODY>	Opens and closes body section of page
<BODY BGCOLOR="color"*>	</BODY>	Sets background color for the entire page
<BODY TEXT="color"*>	</BODY>	Sets text color for the entire page
<BASEFONT SIZE="n"**>	None	Sets the text size for the entire page
<Hn>***	</Hn>***	Designates a heading
<FONT SIZE="n"**>	</FONT>	Changes the font size for a section of text (overrides **BASEFONT**)
<FONT COLOR="color"*>	</FONT>	Changes the font color for a section of text (overrides **TEXT**)
<P>	</P>	Begins and ends paragraph (indicated by line of vertical space)

OPENING TAG	CLOSING TAG	EFFECT
<BLOCKQUOTE>	</BLOCKQUOTE>	Indents text in most browsers; italicizes text in some
 	None	Starts a new text line without adding vertical space
<EM>	</EM>	Italicizes text in most browsers
<STRONG>	</STRONG>	Boldfaces text in most browsers
<I>	</I>	Italicizes text in browsers that recognize tag
<B>	</B>	Boldfaces text in browsers that recognize tag

*For color substitute one of the sixteen predefined colors or a hexadecimal number.
**For n substitute a number between 1 and 7.
***For n substitute a number from 1 (largest heading) to 6 (smallest heading).

WEB SITES

http://bignosebird.com/siteguide.shtml
  Big Nose Bird: a site for beginning and intermediate Web designers, offering a wealth of information about most aspects of the Web

http://www.htmlgoodies.earthweb.com/
  HTML Goodies site from Web designer Joe Burns: lots of tutorials and tips

http://www.inquisitor.com/hex.html
  A site that will convert RGB colors to their nearest hexadecimal equivalents

http://www.lynda.com/hex.html
  Designer Lynda Weinman's explanation of hexadecimal colors, along with a link to a downloadable hexadecimal color palette

http://www.projectcool.com/developer/reference/color-chart.html
  Project Cool's color chart that allows you to see a hexadecimal color as both background and text

http://www.zeldman.com/askdrweb/index.html
  Web designer Jeffrey Zeldman's Ask Dr. Web: very well-designed and authoritative site with lots of good advice

CHAPTER 3

GIFs AND JPEGs AND PNGs, OH MY

Using Images

USING IMAGES

Writing pages with text is relatively simple, but it only scratches the surface of what the Web can do. To really take advantage of the Web's possibilities, you need to include graphics and some page design—and at this point your Web page begins to get more complicated.

Web browsers can read only a few types of graphic formats; that means that any graphic you include on your Web pages must be saved in one of these formats. They must be either GIFs, JPEGs, or PNGs. Explanations of each of these formats follow a little later in this chapter, but let's begin with a little background information.

Some Web Graphic Terms

Clip art.   Copyright-free art available either in print form or electronically on CD-ROM or downloaded from the Web. When you purchase clip art, you purchase the right to use it in publications or on your Web pages. However, some clip-art collections place restrictions on the number of images you can use within a page; be sure to check the fine print on any collection you use.

GIF.   Graphics Interchange Format, a graphics format developed by Unisys Corporation. GIFs are highly compressible and can squeeze down to small sizes. They handle flat colors relatively well and are usually used for drawings and line art.

JPEG.   A format developed by the Joint Photographic Experts Group and used most frequently for photographs on the Web. JPEGs do not handle flat colors well and are not usually used for line art or drawings.

PNG.   Portable Network Graphics, a royalty-free public domain graphic format that supports a variety of advanced features. They are most appropriate for line art and drawings, but are not recognized by browsers earlier than version 4.

Pixel.   A single point on the screen of a graphics-reading computer monitor. *Pixel* stands for *picture element;* on color monitors, each pixel is composed of three dots colored red, green, and blue that converge at some point. The quality of the color monitor depends on the number of pixels it can display and also the number of bits used to represent each pixel, ranging from 8-bit systems (256 colors) to 24-bit systems (more than 16 million colors).

Where Can You Get Web Graphics?

Graphics are available in a variety of formats from many sources. The computer facilities at your school may have CD-ROM clip-art collections for your use, or they may have a scanner you can use to scan in clip art from printed sources such as the collections from Dover Press. If none of these resources are available to you, you can purchase collections from bookstores or computer stores. (Try searching for "clip art" at one of the online bookstores, for example.) You can also create your own art and scan it in, or scan in photographs that you can manipulate in computer graphics programs. Some word processors even have clip-art libraries, although their quality may not be high.

One place to find Web graphics is, not surprisingly, on the Web itself. Several clip-art sites are available, with clip art of varying quality. Clip art, by definition, is available for your use; Web clip art has been provided for Web page authors, usually by the artists, either to publicize their design services or to share their creations with others. If you purchase clip art, either on CD-ROM or in books, you also purchase the right to use it in your publications, either on paper or electronically. (However, some clip-art collections place limits on Web use; be sure to read the fine print regarding permissions.)

You can also download images from Web sites other than clip-art sites, but you must be aware of the copyright issues involved (see the Copyright and the Web box). If the Web site owner–designer is making these images available to you as a gesture of goodwill—that is, if the owner–designer expressly invites you to download the image—then feel free to download, but if the images are there strictly for on-site use (e.g., to augment or illustrate the content of the site), then taking them without permission is equivalent to stealing them. If you then proceed to use these images on your own site, you're violating the copyright of the original owner.

In using clip art from a clip-art site such as Icon Bazaar or Windy's Web, you should credit the source of your graphics on the page itself. Often clip-art sites will provide you with an icon to download and place on your pages that will identify the source of your images. You might also link your site to the clip-art site in order to allow other people to find graphics like yours.

However, one of the best ways to use original and effective images is to create your own. In Chapter 1, a graphics program was mentioned as playing a major part in Web page construction. Either coupled with a scanner or used with an electronic clip-art collection, a graphics program can increase your resources tremendously where images are concerned. Several standard graphics programs are available, including Adobe Photoshop, Macromedia Freehand, and Corel Draw. But the particular graphics program you use is less important than just getting used to a graphics program. Once you become accustomed to one graphics program, others will seem less difficult.

■ COPYRIGHT AND THE WEB ■

It's so easy to take things off the Web—basically point and click—that it's also easy to forget that information and images belong to other people. You can take images and information from the Web for your personal use, but you can't put those images and information on your own Web pages without permission. Look at it this way: you can cut out a picture from a magazine and thumbtack it to your wall: that's personal use. But you can't cut out that picture and put it into *your own* magazine without paying the artist for its use: that would violate the artist's copyright. In the latter case you're using someone else's work for your own financial gain, which doesn't constitute fair use. Here are some guidelines to keep in mind:

- Don't copy material from another site without permission, even if you give credit to the author.
- Don't use someone else's picture, voice, performance, and so forth without permission, and get that permission in writing.
- Don't assume that material posted on someone else's Web site belongs to the person who posted it. (The author may be violating someone else's copyright without your knowledge.)
- Quote briefly, or use a paraphrase of someone else's words, always being sure to give full credit to your source.
- Give credit to the sources of your information, both on your Web page itself and through a link to the original source.
- If you're mounting your pages on your school's Web server, be sure to check any campus policies on copyright and fair use.

■ TRY IT

Go to one of the standard Web search engines (e.g., Northern Light, Yahoo!, Excite, AltaVista, etc.), and locate some clip-art sites. Browse through the selection and see what's available. Bring the URL of the best site you found to class with you.

PLACING GRAPHICS ON YOUR PAGES

Before we start looking at the ins and outs of creating graphics, there's one more point to emphasize: these images aren't really placed on your pages. The tag for placing graphics (`IMG SRC`—which we discuss shortly) is actually a direction to the browser. In effect, the tag tells the browser to go to the server and request a particular graphic file at this particular point on the page. Once the server supplies the graphic file, the browser will display it at the point on the page where the tag appears.

Why is it important to keep this in mind? It's easy to forget that the graphic isn't embedded on the page just because you wrote a tag that directs the browser to put the graphic there, particularly if you're using a WYSIWYG graphic-interface authoring program that shows you the graphic as if it were part of the page (yet another good reason to learn HTML for yourself). Also, if you're accustomed to using page-layout programs such as Pagemaker or Quark Xpress (which do embed graphics on the page), it's hard to remember that Web pages are different. However, to make that image appear at that point on the page, you need to supply the graphic file along with the HTML file for the page. If you're uploading the files to a Web server, you'll need to upload the graphic file along with the page file; if you're handing in your Web page on disk, you'll need to make sure that the graphic files are included along with your page files. Otherwise, when the browser opens your page, your reader will see the dreaded missing image icon (Figure 3.1).

IMAGE FORMATS

Graphics can be seen by browsers only if they're saved in one of three formats: GIF, JPEG, and PNG.

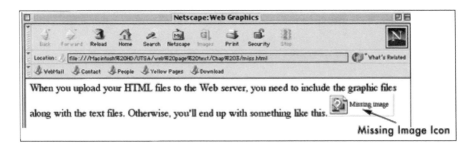

FIGURE 3.1   The Missing Image

GIF

GIF stands for Graphics Interchange Format; it's a graphics format developed by Unisys Corporation. GIFs were developed to be highly compressible images; that means that you can squeeze a large image down to a much smaller size if you convert it to a GIF. Moreover, GIF images are "lossless"; that is, when you decompress your image in the browser, it will look the same as the original. GIFs also can mask one color, which means that you can create transparent backgrounds and place your graphic directly against the background of your page. (Actually you're just covering up the background color of your graphic.) You also can use GIFs to create the equivalent of simple flip-book animations, putting several GIFs in a loop by using a GIF animation program.

One drawback of GIFs is that they can use only 256 colors; that may seem like a lot, but in fact those 256 colors leave out quite a few. Most photographs will suffer if you save them in GIF format because you won't be able to capture the subtle gradations in color from one part of the photograph to another. For this reason, GIFs are used most frequently for line art: cartoons, drawings, sketches, and so forth. Photographs are better saved as JPEGs, unless the photograph is a small one where the loss of detail won't matter and the small file size is a major consideration.

JPEG

JPEG (a format developed by the Joint Photographic Experts Group) is the format used most frequently for photographs on the Web. JPEGs deal with the brightness and color of photographs separately, compressing the more subtle gradations in color that all photographs have. It's a "lossy" process, which means it loses information in compression—JPEG images won't have the same quality as the original image. JPEGs can be saved at various levels of quality (usually, low, medium, high, and maximum), but the better the quality, the sharper the picture—and the larger the file size. In general, with Web graphics, you'll try for the lowest quality setting you can get by with, because you'll want to keep your file size as small as you reasonably can. (The larger the graphic file size, the longer it will take the Web browser to download it.)

JPEGs don't handle flat colors well; the process is designed to work with graduated color. Thus for images with flat color, such as drawings or small-size images, GIFs are a better choice because they handle flat color better and are more compressible.

PNG

Portable Network Graphics (PNG) is a royalty-free public domain graphic format that supports a variety of advanced features. PNG graphics include more color and gray-scale shades than do GIFs and can be compressed into much smaller files without sacrificing image quality. PNG also includes a better method for creating transparent backgrounds than does GIF, as well as other, more advanced Web-writing features. So why isn't everyone using PNG instead of GIF? At the moment, the format isn't supported by any browsers earlier than version 4 of Netscape and Internet Explorer, and these browsers don't really support the format effectively. Moreover, graphics programs have been slow to adopt PNG as a possible format for saving graphics. So PNG is more the format of the future than the present. Still, you need to know that PNG exists; one of these days it will probably supersede GIF.

Which Format to Choose

To sum up

IF...	THEN...
Your graphic is a photograph	Use JPEG
Your graphic is a line drawing	Use GIF
Your graphic includes areas of solid color	Use GIF
Your graphic includes more than 256 colors (and you don't want to reduce the number)	Use JPEG
You want a transparent background in your graphic	Use GIF
You want to animate your graphic	Use GIF

■ FILE SIZE ON THE WEB ■

You'll find that we keep coming back to this issue of file size, particularly with your graphics. Look at it this way: on average, considering modem speed, your Internet connection, the Web-server speed, and the like, every kilobyte (K) of file size requires around one second to download. That's why many Web designers caution that no page should be larger than 70K total. A user who's interested may be willing to wait for a minute while your page downloads. But any longer than that may make a user surf on to faster pages. Test yourself sometime: how fast do your favorite Web pages download on your modem connection? How long are you willing to wait for a page to be completely visible before you feel like hitting the stop button?

Converting Your Graphic

If you've scanned a graphic, it may be a TIFF or a PICT or a BMP or an EPS or another graphic format; electronic clip art may also be in any one of these various forms. Your first step should be to determine the current format of your graphic. You can look at the graphic on your desktop in either Mac or Windows to see what the format is, or you can open the graphic in your graphics program and check it there. Different graphics programs use different processes to convert images from one format to another. Consult the documentation on your graphics program to find out how to do it.

Converting a JPEG.   When you choose JPEG as the format in which you want to save a graphic, you'll probably be given a choice of quality levels (usually low, medium, high, and maximum). If you want to experiment to see how low you can make the quality level and still get an acceptable graphic, try saving your graphic at various levels under different names (e.g., *imagelow, imagemed, imagehigh, imagemax*), and then look at the different versions side-by-side. Remember, you want a graphic that doesn't take up much memory, but not one that's hard to recognize.

▪ GRAPHIC RESOLUTION ▪

Resolution refers to the sharpness of an image: a high-resolution image will be very sharp, whereas a low-resolution image may be slightly blurry. When you scan an image for print, you usually scan at a relatively high resolution, frequently 200 dpi (dots per inch) to 300 dpi. However, the resolution of most computer monitors is only 72 dpi. Ultimately, you'll want to save your image at 72 dpi (which will result in a smaller file size); however, if you scan it at a higher resolution than that and then reduce it, you'll have a high-quality version of your image to start with. (It's easier to maintain quality while reducing than while enlarging.) Scan in your image at 200 dpi, then save another version of it in your graphics program at 72 dpi so that you can retain both the high- and low-resolution versions; the result should reproduce an image that has both a high level of quality and a smaller file size.

Saving Your Graphic

Once you've converted your graphic to GIF or JPEG (or PNG if you wish), you should save it with an extension that indicates the type of graphic it is. (Extensions are the letters at the end of a file name that tell the browser the type of information the file includes.) The extensions are `.gif`, `.jpg`, `.png`. The complete graphic name would look like this: `image.gif`, `image.jpg`, `image.png`. Incidentally, the rules for naming and saving your files also apply to graphic files—keep the file name under eight characters and don't use any spaces. These extensions tell the browser what type of graphic you're using so that the browser knows how to interpret the file.

There's one important point to keep in mind, however: *simply attaching the extension to the graphic name will not convert the graphic to another mode.* In other words, if you haven't converted your graphic to a GIF in your graphics program, simply adding `.gif` to the graphic name won't do it.

You must also use the right extension for the format of your graphic. If your graphic is a JPEG and you forget and name it with `.gif`, not only will it not be converted to a GIF, it won't show up in the browser!

■ TRY IT

Either from one of the clip-art sites you located in the previous exercise or from another source (e.g., CD-ROM clip-art collections, clip-art books, etc.), locate at least three images that you like. Make sure that they're either GIFs, JPEGs, or PNGs. If they're not, convert them to the appropriate format in your graphics program.

PLACING YOUR GRAPHIC ON YOUR WEB PAGE

The basic tag for inserting an image on your page is `IMG SRC`. `IMG`, which stands for "image," is the tag; `SRC`, which stands for "source," is the attribute; and the name of the image is the value. The whole tag looks like this (there's no closing tag, because this isn't a container):

```
<IMG SRC="idea.gif">
```

Placing a Graphic

1. Prepare your image in advance and put it in the folder with the other contents of your page (e.g., the HTML file). Remember: only GIF, JPEG, or PNG images can be used on Web pages.
2. On your HTML page, place the cursor where you want the image to appear
3. Type `<IMG SRC="image">` (Instead of `"image"`, use the name and type of your image, e.g., `idea.gif`)

The `IMG SRC` tag can be used with all graphic types: GIF, JPEG, and PNG. It can also be used to place GIF animations.

■ THE RIGHT NAME ■

Sometimes you'll create an `IMG SRC` tag only to have the dreaded missing image icon appear on your page—even though the image is in the folder and is in the right format. The first thing to look at when this happens is the way you typed the name of the graphic in the `IMG  SRC` tag. Check your spelling, to begin with: did you spell the name of the graphic the same way in your `IMG SRC` tag as you did when you saved it? Next, check your capitalization. Browsers are case sensitive: that means that if you capitalized the name of the image when you saved it in your graphics program (e.g., `Idea.gif`) but typed it lowercase in your HTML (e.g., `idea.gif`), it won't show up on your page. Make sure that you type the graphic name in your HTML *exactly* the same way you did when you saved it. The same thing goes for the extension: browsers will read both `idea.GIF` and `idea.gif`, but as far as the browser is concerned, they're two different graphics. It's a good idea always to type the extension in the same way, that is, either *always* capitalize it or *never* capitalize it.

Adding Alternate Text

Occasionally, Web browsers won't show graphics. Some older browsers don't show them at all, and some users turn the graphics off to increase the speed of their browsers. Also, visually impaired users may employ browsers that "read" the graphic tags as text. For all of these users, you can insert the `ALT` tag, which offers a text version of the graphic. (`ALT` tags are also helpful for users with slow connections; the `ALT` tag provides users with some idea of what the graphic is that they're waiting to download.) You provide the `ALT` version of the graphic within the graphic tag, so that it looks like this:

```
<IMG SRC="idea.gif" ALT="A better idea">
```

Adding an Alternate (Text) Version of a Graphic

1. Place the cursor where you want the image to appear.
2. Type `<IMG SRC="image" ALT="text"` (Instead of `"text"`, type the text that should appear if the image does not; e.g., `<IMG  SRC="rose .gif" ALT="red rose">`)

Your HTML should look something like this:

```
<HTML>
<HEAD>
     <TITLE>Images 1</TITLE>
</HEAD>
<BODY BGCOLOR="white" TEXT="#3300FF">
<P><FONT SIZE="4">The first time I tried using
graphics on my Web page I had a terrible time. I
```

was using a graphic-interface editor and I didn't
understand HTML too well. I wanted my text to
wrap around the graphic, but I didn't know how to
do it. So I ended up with this.</P>
<IMG SRC="sour.gif" ALT="Sour-Faced Gent">
<P>It wasn't what I really wanted</P></FONT>
</BODY>
</HTML>

Figure 3.2 shows what this looks like in a browser.

■ TRY IT

Write the HTML and text—and supply the graphic—for a page that includes
the following:

1. A paragraph
2. A graphic (either GIF, JPEG, or PNG), with the ALT attribute

   Remember to include the basic template: <HTML></HTML>,
   <HEAD></HEAD>, <TITLE></TITLE>, <BODY></BODY>

FIGURE 3.2   Page with One Graphic

If your instructor agrees, you can go on working with the Web page you created in the previous chapter, or you can begin a new page on another subject. (See the writing assignment at the end of this chapter.)

Aligning an Image

You can align an image with a line of text or you can wrap a longer text section around an image. Let's start with aligning first because it's a little easier.

Aligning an Image

1. Type `<IMG SRC="image"` (For `"image"`, substitute the name of your graphic.)
2. Type `ALIGN= top, middle,` *or* `bottom>`
3. Type the text to be aligned with the image.

Note: Be careful with multiple images on the same line; the `ALIGN` command will have different effects, depending on which image is first and which is taller.

If you use *top,* you'll align the image with the highest element in the line; *middle* aligns the middle of the image with the baseline of the text, and *bottom* aligns the bottom of the image with the bottom of the line of text. Your code for aligning an image will look something like this:

```
<HTML>
<HEAD>
        <TITLE>Images 1</TITLE>
</HEAD>
<BODY BGCOLOR="white" TEXT="#3300FF"> <BASEFONT
SIZE="4">
<P>The first time I tried using graphics on my
Web page I had a terrible time. I was using a
graphic-interface editor and I didn't understand
HTML too well. I wanted my text to wrap around
the graphic, but I didn't know how to do it. So I
ended up with this.</P>
<IMG SRC="sour.gif" ALT="Sour-Faced Gent"
ALIGN="middle">
It wasn't what I really wanted
</BODY>
</HTML>
```

In a browser it will look like Figure 3.3.

The first time I tried using graphics on my Web page I had a terrible time. I was using a graphic-interface editor and I didn't understand HTML too well. I wanted my text to wrap around the graphic, but I didn't know how to do it. So I ended up with this.

It wasn't what I really wanted

FIGURE 3.3   Middle Alignment

Although both alignment and text wrap use the **ALIGN** command, you can't align an image with text and wrap text around it at the same time. Text wrap is a little more complicated.

■ TRY IT

Use the same page you worked with in the last exercise, but this time align the text with the graphic.

Wrapping Text

When you wrap text around an image, you place the two side-by-side and flow the text around the graphic. For this to work well, you need a graphic and text block that are about the same size. A large graphic with a small text block will end up looking like the alignment in Figure 3.3. Always place the image *before* the text you want to wrap around it, then indicate in the tag whether you want the image on the left (text on the right) or the image on the right (text on the left).

Wrapping Text around an Image

1. Type `<IMG SRC="image"` (For `"image"`, substitute the name of your graphic.)
2. Type `ALIGN=LEFT>` to align the image to the left of the screen (text flows to the right) *or* `ALIGN=RIGHT>` to align the image to the right edge of the screen (text flows to the left).
3. Type the text that should flow around the image.

Your HTML should look something like this:

```
<HTML>
<HEAD>
        <TITLE>Image 2</TITLE>
</HEAD>
<BODY BGCOLOR="white" TEXT="#FF0000"> <BASEFONT
SIZE="4">
<IMG SRC="idea.gif" ALT="Bright Idea"
ALIGN="left"> After  a while I started to get the
hang of text wrap. The hardest thing for me was
getting the right amount of text so that my text
would wrap around my images effectively. I was
accustomed to page layout software and I wanted
my Web pages to look as "classy" as my hard copy
did. When I did text pages, I could make fine
adjustments in the font size or even in the font
itself (e.g., Palatino is bigger than Times;
Bookman is thicker than Palatino). At first with
Web pages, all I could adjust was the font size—
and if I made the text <EM>too</EM> big on my Mac
(which has slightly bigger text sizes), it would
look really crummy on my PC. Until I learned how
to use tables for Web page layout, I was not a
happy camper.
</BODY>
</HTML>
```

Figure 3.4 shows how it looks in a browser.

You can also wrap text between two images, so that you have the text flowing down the middle like a river. However, at some point, you'll need to end the text wrap so that the images don't go on affecting your text: an image will push text away until either a line break or the CLEAR tag. When you reach the end of the section you want to wrap around the graphic or between the two graphics, insert BR CLEAR= and then use either left, right, or all (depending on where the graphic was placed) so that you can turn off the text wrap.

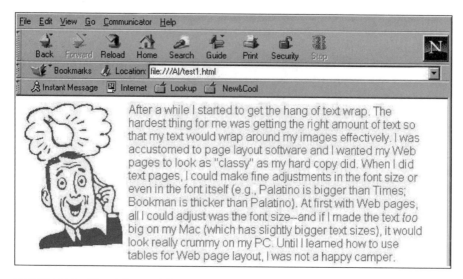

FIGURE 3.4 **Text Wrap with One Image**

Wrapping Text between Two Images

1. Type <IMG SRC="image" (For "image", substitute the name of your graphic) ALIGN=RIGHT [or LEFT]>
2. Type the text that should flow around this image.
3. Type <IMG SRC="image" (For "image", substitute the name of your graphic) ALIGN=LEFT [or RIGHT]>
4. Type <P> to begin a new paragraph aligned with the last image.
5. Type the text that should flow around the second image.
6. Type <BR CLEAR=all>

Your HTML will look something like this:

```
<HTML>
<HEAD>
        <TITLE>Image 2</TITLE>
</HEAD>
<BODY BGCOLOR="white" TEXT="#FF0000"> <BASEFONT
SIZE="4">
<IMG SRC="idea.gif" ALT="Bright Idea"
ALIGN="left"> After a while I started to get the
hang of text wrap. The hardest thing for me was
getting the right amount of text so that my text
```

would wrap around my images effectively. I was accustomed to page layout software, and I wanted my Web pages to look as "classy" as my hard copy did.

```
<IMG SRC="prize.gif" ALT="First Prize"
ALIGN="right"><P>When I did text pages, I could
make fine adjustments in the font size or even in
the font itself (e.g., Palatino is bigger than
Times; Bookman is thicker than Palatino). At
first with Web pages, all I could adjust was the
font size—and if I made the text <EM>too</EM> big
on my Mac (which has slightly bigger text sizes),
it would look really crummy on my PC.</P>

<P>Until I learned how to use tables for Web page
layout, I was not a happy camper because nothing
I did as a Web author looked exactly the way I
wanted it to look.</P><BR CLEAR="all">
</BODY>
</HTML>
```

In a browser it looks like Figure 3.5.

■ TRY IT

Write the HTML and text for a page that includes text wrap around a single image. Remember to include the basic template: <HTML> </HTML>, <HEAD></HEAD>, <TITLE></TITLE>, <BODY></BODY>

You can go on using the page you've been working with in previous exercises if you wish.

Putting Space around Your Graphics

Once you place your graphics, you may feel that the text comes a little close, particularly if you have a small graphic and a lot of text. You can add some space around your graphics by using the **VSPACE** and **HSPACE** attributes. **VSPACE** stands for vertical space, the space at the top and bottom of your image; **HSPACE,** as you might guess, is horizontal space, space to the right and left of your image. To put space around your image, you'd type **VSPACE=** and/ or **HSPACE=** and then add a number of pixels, the amount of space you want added to your graphic.

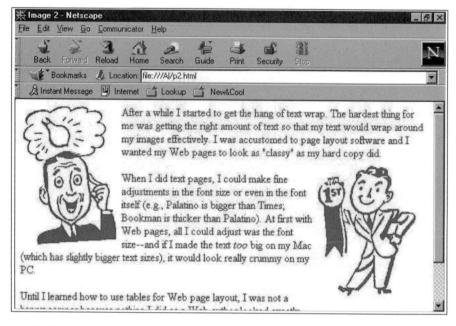

FIGURE 3.5   Text Wrap with Two Pictures

Putting Space around Your Graphic

1. Type `<IMG SRC="image"` (For `"image"`, substitute the name of your graphic.)
2. Type `HSPACE=x` (For `x`, substitute the number of pixels of space you want added at both the right and left sides of the image.)
3. Type `VSPACE=y>` (For `y`, substitute the number of pixels of space you want added at both the top and bottom of the image.)

Note: You can have just `HSPACE` or `VSPACE`; you don't have to include both.

Your HTML will look something like this:

```
<HTML>
<HEAD>
        <TITLE>Image 2</TITLE>
</HEAD>
<BODY BGCOLOR="white" TEXT="#FF0000"> <BASEFONT
SIZE="4">
```

```
<IMG SRC="idea.gif" ALT="Bright Idea"
ALIGN="left" HSPACE="10" VSPACE="10"> After a
while I started to get the hang of text wrap. The
hardest thing for me was getting the right amount
of text so that my text would wrap around my
images effectively. I was accustomed to page
layout software and I wanted my Web pages to look
as "classy" as my hard copy did.
<IMG SRC="prize.gif" ALT="First Prize"
ALIGN="right" HSPACE="10" VSPACE="10"><P>When I
did text pages, I could make fine adjustments in
the font size or even in the font itself (e.g.,
Palatino is bigger than Times; Bookman is thicker
than Palatino). At first with Web pages, all I
could adjust was the font size—and if I made the
text <EM>too</EM> big on my Mac (which has
slightly bigger text sizes), it would look really
crummy on my PC.</P>
<P>Until I learned how to use tables for Web page
layout, I was not a happy camper because nothing
I did as a Web author looked exactly the way I
wanted it to look.</P><BR CLEAR="all">
</BODY>
</HTML>
```

Figure 3.6 shows how this looks in the browser.

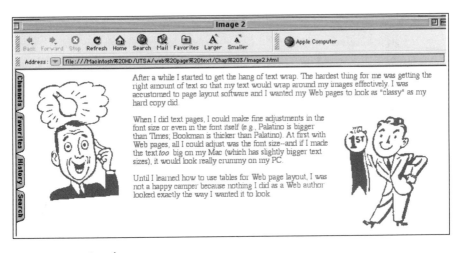

FIGURE 3.6   Spacing

▓ TRY IT

Write the HTML and text for a page that wraps text between two images. Include HSPACE and VSPACE. Remember to include the basic template: <HTML></HTML>, <HEAD></HEAD>, <TITLE></TITLE>, <BODY></BODY>

You can use the page you worked with in the previous exercise if you wish.

BACKGROUND TILES

In Chapter 2, you learned how to change the color of your background from the default gray or white. You can also place images in the background of your page, such as wallpaper on your computer's desktop or watermark pictures in the background of a text page. These images are "tiled"; that is, they're repeated horizontally and/or vertically, depending on the size and shape of the image. The same rules hold true for background images as for other images, however: they must be either GIFs, JPEGs, or PNGs, and they *must* be small files, in terms of both their memory requirements and their dimensions.

Background Tile Designs

Your main consideration when you create a background tile is making sure the combination of tile and text will still be readable. There are several points to keep in mind:

- If you use a light background, keep your text dark, and vice versa (i.e., dark backgrounds need light text).
- If you use a dark background tile (or color), be sure to change *all* the text color settings, including the settings for link colors (you don't want the default dark blue and dark purple links to disappear against your background).
- With patterned background tiles, try to use patterns that are either all dark or all light colors. Otherwise, your type will show up one way on one part of the pattern and another way on another part.
- Avoid "busy" background patterns that interfere with the type. Test your background tile and type to make sure that the combination works.

▓ SCREEN SIZE ▓

The average size for browser windows is usually given as 640 pixels by 480 pixels. That's actually a bit smaller than the way browsers appear on some monitors, but it's a bit larger than others. If you're scaling an image for background, you can use 640 pixels as your basic width. That way, if you want a long narrow background tile, you can make it 1,200 pixels wide and be fairly certain it will repeat only vertically.

Creating a Background Tile

You can create background tiles of various sizes, depending on how you want to fill your screen. However, your image must be either a square or a rectangle. The size of your tile will determine how often it repeats on the screen, and whether it will repeat both horizontally and vertically or only vertically. However, there's good reason to keep the dimensions of your background tile relatively small: the bigger the tile, the larger the memory requirements and the longer the download. Remember: the size of your background tile will be added to the total for your page. If you have a 30K background tile along with two 10K graphics, the total will be 50K for the page. This page will take about fifty seconds to download, considering modem, connection, and server speed.

Keep in mind that, like wallpaper, a tile with a noticeable pattern may not match up seamlessly in the background: the more obvious the border around the edges of your tile, the more evident the tiling effect will be on your page. That's not necessarily bad: maybe you want to emphasize the effect of a repeated pattern. Yet if you'd rather not emphasize it, create a background tile without an obvious border. One easy way to create a background pattern is simply to select part of another graphic (e.g., clouds, ironwork, rippling sand, etc.).

Creating a Square Background Tile by Copying a Pattern

1. In a graphic program, open the graphic you want to use as the source of your background tile.
2. Select an area about fifty pixels by fifty pixels.
3. Copy it.
4. Close the graphic and open a new file sized about fifty pixels by fifty pixels.
5. Paste the small section of the graphic you copied into the new file.
6. Save your tile and then convert it to a GIF (or save the file as a JPEG or PNG if another format is more appropriate).

Another way to create an effective background tile is to use a long, thin horizontal stripe (see Figure 3.7). If you use a pattern or a contrasting color at the left end of the stripe and leave the rest of the stripe a solid color such as white or cream, you can have a narrow column using a background texture going down the left side of your page, and your text will show up effectively against the solid color on the right.

FIGURE 3.7   A Stripe Tile

If you make the stripe about 1,200 to 1,400 pixels long, it will show up only once horizontally while it repeats vertically. The effect will be a vertical stripe or column running down the left side of your page (see Figure 3.8).

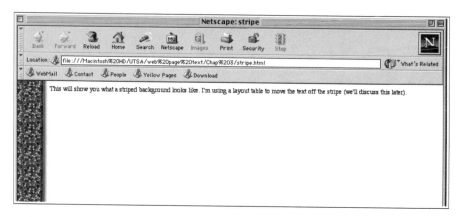

FIGURE 3.8   Stripe in Use

■ TRY IT

Create a background tile in your graphics program. You can use either a square tile (say 50 by 50 pixels) or a long horizontal tile (say 25 pixels high by 1,400 pixels wide).

Writing the HTML

You place your background tile as an attribute and value inside the BODY tag. BACKGROUND is the attribute, whereas the name of the graphic is the value. The entire tag looks like this:

```
<BODY BACKGROUND="tile.gif">
```

Your HTML should look something like this:

```
<HTML>
<HEAD>
      <TITLE>Image 2</TITLE>
</HEAD>
<BODY BACKGROUND="ttile.gif" TEXT="#FF0000">
<BASEFONT SIZE="4">
<IMG SRC="idea.gif" ALT="Bright Idea"
ALIGN="left" HSPACE="10" VSPACE="10"> After a
while I started to get the hang of text wrap. The
hardest thing for me was getting the right amount
of text so that my text would wrap around my
images effectively. I was accustomed to page
layout software and I wanted my Web pages to look
as "classy" as my hard copy did.
<IMG SRC="prize.gif" ALT="First Prize"
ALIGN="right" HSPACE="10" VSPACE="10"><P>When
I did text pages, I could make fine adjustments
in the font size or even in the font itself
(e.g., Palatino is bigger than Times; Bookman is
thicker than Palatino). At first with Web pages,
all I could adjust was the font size—and if I
made the text <EM>too</EM> big on my Mac (which
has slightly bigger text sizes), it would look
really crummy on my PC.</P>
<P>Until I learned how to use tables for Web page
layout, I was not a happy camper because nothing
I did as a Web author looked exactly the way I
wanted it to look.</P><BR CLEAR="all">
</BODY>
</HTML>
```

Figure 3.9 shows what it looks like in the browser.

■ TRY IT

Write the HTML and text for a page that includes a background tile. Include at least a paragraph of text and another graphic in addition to the tile. You can go on working with the page you created in previous exercises if you wish.

You have enough HTML now to put together a simple Web page, using text and graphics; your text will wrap, your graphics will show up, and the results will probably be effective. Let's step back now and talk more generally about creating your pages, the topic of Chapter 4.

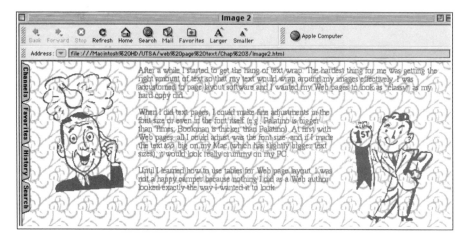

FIGURE 3.9   Page with Background Tile

■ WRITING ASSIGNMENT

Your writing assignment for this chapter will get a little more complicated. This time you'll write *four* paragraphs (`<P></P>`) about a subject of your choosing, but this time you'll also include *two* images of some kind. Your topic can be different from your first page, but you could stay on the subject of computers and Web surfing. You could write about what kind of design attracts you when you're on the Web, for example, or what irritates you most about ineffective Web pages. If you wish (and if your instructor approves), you can continue to work with the page you created for Chapter 2, expanding and revising to include images as part of your presentation.

This page should have the following:

- Text wrapped around a single image
- Text wrapped *between* two images
- A changed background (using either background color or a background tile)
- Colored text

Tips. Try to choose images that fit your topic and that complement each other. If you use clip art, make sure all the images use the same style—don't use one very modern one and one that looks old-fashioned, for example. If you're writing a personal experience essay, you can even use your own photographs (scanned in and turned into JPEGs, of course). Remember: anything on a page that isn't text will draw your readers' attention first. Look for images that are worth the attention they'll get.

TAGS USED IN CHAPTER 3

OPENING TAG	CLOSING TAG	EFFECT
`<IMG SRC="image.gif"*>`	None	Tells browser to insert a graphic file
`<IMG SRC="image.gif"*` `ALT="Alternate text">`	None	Inserts text description of graphic for browsers that are not graphics-enabled

`<IMG SRC="image.gif"*`
`ALT="Alternate text"`
`ALIGN="top" "middle"`
or `"bottom"**>`
	None	Aligns graphic with a single line of text

`<IMG SRC="image.gif"*`
`ALT="Alternate text"`
`ALIGN="left"` *or* `"right"**>`
	None	Wraps text around a graphic

`<BR CLEAR="left" "right"` *or* `"all">**`	None	Turns off text wrap

`<IMG SRC="image.gif"*`
`ALT="Alternate text"`
`VSPACE=n*** HSPACE=n***>`
	None	Adds vertical and/or horizontal space around a graphic

`<BODY BACKGROUND="image.gif"*>`	`</BODY>`	Inserts a repeating background tile

*Insert the name and type of your graphic file.
**Choose *one* of these alternatives.
***Insert a number of pixels.

WEB SITES

http://bignosebird.com/losewait.shtml
Big Nose Bird's graphic tutorial, stressing how to manage the size of your graphics

http://home.netscape.com/assist/net_sites/impact_docs/index.html
Netscape's words of wisdom about using graphics

http://megawebtools.com/
Mega Web Tools: a directory of sites offering free clip art, animations, sounds, and so on

http://webbuilder.netscape.com/Business/Law/ss08.html
Discussion of legal issues in using artwork online

http://netcenterbu.builder.com/Graphics/Graphics101/
Netscape's tutorial on Web graphics

http://www.libpng.org/pub/png/
More information on the PNG format

http://www.mediabuilder.com/
Media Builder site featuring clip art, backgrounds, gif animations, and the like

http://www.netscape.com/assist/net_sites/bg/backgrounds.html
Selection of background tiles at Netscape

http://www.pambytes.com/
Pam Bytes: original clip art and background tiles from a Web designer

http://www.windyweb.com/design/
Windy's Web: another designer's clip-art site, with interesting background tiles

http://www.zeldman.com/abtgraph.html
Dr. Web's Graphics Clinic

Finally, some government sources for copyright-free photographs:

http://info.fws.gov/images.html
National Fish and Wildlife Service

http://nix.nasa.gov/
NASA Image Exchange

http://www.nasa.gov/gallery/index.html
NASA on the Web, a linked list of Web-based resources available through NASA

http://www.photolib.noaa.gov/lb_images/searchit.html
National Oceanic and Atmospheric Administration

CHAPTER 4

HOW SHOULD I SAY IT
Writing for the Web

FROM INSTRUMENT TO THEORY

If you've finished the first three chapters of this book, you've been given a quick introduction to the basics of HTML. You know now how to write tags and how to create most parts of a simple Web page. But although you know how, you may not know what. What should go into a Web page?

In many ways the Web, like a lot of new media, has a certain wide open quality. There's a sense that the Web has fewer rules, that you can do more there. Whereas that's certainly true in terms of elements you can include, it's a mistake to think there are no rules for writing on the Web. The rules (or, rather, the standards) have grown out of experience, as people move around the Web and discover what works in what context for what purpose. A lot of ineffective pages are still on the Web, but their problems frequently come from not knowing what makes a page effective.

PURPOSE AND AUDIENCE ON THE WEB

So what should you do when you start writing a Web page? Basically do the same thing you do when you start writing any kind of page: consider your purpose and your audience. But let's admit up front that figuring out your purpose, and particularly your audience, may be more challenging for a Web page than for, say, a letter to the editor of your local newspaper.

Purpose

The idea behind purpose is simple: what are you trying to accomplish with this site? You might have a variety of purposes that will influence the kind of writing and design you do on your site. Determining your purpose is part of a larger question—why do you want to write a Web page in the first place? Are you sharing your enthusiasm for a topic? Are you passing on your concern about an issue? Would you like to interest others in something you've made or done?

57

One way to begin thinking about your purpose is to ask yourself what keywords a user might use to find your page on the Web. Go to one of the major search engines, such as Northern Light, Yahoo!, Excite, Infoseek, or AltaVista, and look at the major topics and subtopics it lists. Where would you want your page to be listed and why?

In its Web style guide, Sun Microsystems suggests four common purposes that may help you to think about your pages: service, sales, information, and link lists. To these we add a fifth category: pleasure.

Service.   You may be providing a service for your visitors: maybe you're hosting an electronic bulletin board or perhaps you're furnishing a catalog of products with a form for ordering them online. What kind of site would you construct?

Consider, first of all, that your visitors will probably want to find the service they need quickly. Thus your pages will probably be short and organized for quick scanning: lots of headings, perhaps bulleted lists (see Chapter 7 for information on the HTML tags for lists), anything that will help readers to navigate through your information quickly. Most of all, a site like this will avoid clutter: no unnecessary graphics or anything else that might increase the time needed to download the pages.

Sales.   The purpose for your site might also be promotional—you have a product or service to push, and readers are potential customers. Although a site like this might actually be selling a product, it might also be selling something less tangible, such as a club function you'd like the public to buy tickets for or a social problem you want to alert your visitors about. Of course, when you put your résumé or "Webfolio" on the Web, you're actually selling yourself!

Your readers here are still likely to value clear organization, although they may be less specific about what they're looking for than are the visitors at a service site. They still won't have much patience with extended text sections or organization that hides the vital information. Here again, your site should be set up to present the information quickly and forcefully. Graphics will play a more important part; if possible, let pictures do part of your talking so that they can catch the reader's attention. Above all, don't make your visitors search for information; they may not bother.

Information.   Informational sites, like those from government agencies or organizations such as the World Wide Web Consortium, can assume a slightly more committed visitor than service or sales sites (magazine and newspaper sites are also in this category). Because these readers need or want the information that a site presents, they're likely to be more patient.

Pages on informational sites are frequently longer and more detailed than those on sales sites. In fact, given the amount of information necessary to cover certain topics, shorter pages can actually be more frustrating on this kind of site. However, that doesn't mean that these sites don't require clear organization and design, or that you can simply post pages of text without adapting them for the Web. People who need information are also interested in speed and efficiency, and you never want visitors to feel as if their time is being wasted. Once again, you'll use headings and other attention getters to help your readers scan through your pages.

Link Lists.   *The Sun Microsystems Guide to Web Style* also discusses a particular kind of informational site: one devoted to a linked reference list. Your purpose at such a site is a combination of service and information: you're providing a portal to a series of sites that will supply more information. Such sites are relatively easy to construct because they're usually collections of other site names and URLs. However, an effective link list will be structured and organized for easy reading. It will include annotations and may be categorized using headings and subheads.

Pleasure.   Finally, a fifth category to add to the four listed in the Sun Web style guide is pleasure. Some sites are principally set up to provide a pleasurable experience—entertainment and delight at the creativity of the site designers. Bali Highway (http://www.balihighway.com/), for example, is a multimedia site with a variety of intriguing, experimental pages. There is also a variety of experimental e-zines that work with creative uses of hypertext, such as *Kalibre 10K* from Denmark (http://www.k10k.net/) or *Impossible Object* from Brown University (http://www.brown.edu/Departments/English/ Writing/object/index.html). The programming and design that go into these sites are probably beyond your skills as a beginner, but they certainly can give you something to shoot for!

■ TRY IT

Do some Web surfing now—check on some of your favorite sites or look at some new ones. (Yahoo's Picks of the Week—http://www.yahoo.com/picks/— are a fun place to start.) What kind of purposes are these sites designed to accomplish? Can you categorize their purposes as service, sales, information, link lists, or pleasure? Bring the URL of your favorite site to class to share.

Audience

If you ask yourself who the audience for your Web page is, you may be tempted to answer "everyone who owns a computer with an Internet connection." But that's not really true. No one visits every site that's out there; your readers will be coming to you for a particular reason. Maybe they're interested in your information. They may want to buy your product. Perhaps they need to get something that only your site has (such as directions to the soccer field). They may be teachers and/or employers evaluating your site. Or maybe they're your friends and relations, checking out what you've been spending your time on.

Before you begin to write, you need to figure out what the average visitor will be looking for on your site, what the purpose will be in visiting you. Then you need to figure out how to accommodate your visitors while accomplishing your own purpose in creating the site. As we've emphasized repeatedly, most Web readers have limited time and patience. If they can't find what they're looking for with minimal stress, they may move on quickly.

Web designer Jeffrey Zeldman has developed a useful characterization of Web audience types (in "Design Your Audience," http://www.alistapart.com/stories/who/); he divides them into users, viewers, and readers.

Users.   Users, according to Zeldman, are trying to accomplish a task. The Web is a means to an end for users; they want to find something or write something or design something. Because users are looking for something specific, sites designed for them need to be clear and easy to navigate; they should avoid complex organization and slow-loading graphics, no matter how spectacular.

Most search engines (such as Yahoo! or Excite) are examples of sites written for users. You come to them looking for specific information, and you want to find it quickly. An example of a user site is in Figure 4.1.

This page for the National Weather Service offers visitors a number of menus and other ways to approach information. It's essentially concerned

FIGURE 4.1   A User Site

with directing users on to particular weather data and doing so quickly and efficiently.

Viewers.   Viewers, on the other hand, want entertainment. For them, each site is an experience in itself, and viewers are willing to go along for the ride. The classic viewer is the game player: gamers expect lots of images, intriguing activities, even "bleeding-edge" multimedia. If it's viewers you want at your site, be prepared to spend a lot of time developing clever and innovative design.

Graphic design sites are frequently viewer sites. Try Counterspace or T26 fonts or The Graphic Home. Figure 4.2 shows the opening page for a viewer site.

Figure 4.2 comes from *Quokkasports,* an online sports magazine with a variety of Shockwave movies and other effects. It's meant to provide the reader with an experience along with a series of articles to read.

Readers.   Readers, according to Zeldman, are "somewhere between users and viewers." Readers not only want to read but also want to view. They look for readable font sizes and clearly written text, but they also want attractive, elegant pages that will provide visual pleasure. Readers aren't necessarily using your Web site as a means to an end or as a thrill ride; they're interested in what you have to say, but they want to enjoy the experience too.

Many of the magazines on the Web qualify as reader sites, full of text but also full of design. Check out *Salon* or *Feed* or *Word*—all are magazines that appeal to Web readers. In Figure 4.3 you can see the opening page for a reader site.

The Legends site offers texts of a variety of legends and folktales, but it's also attractively designed with visually appealing pages.

FIGURE 4.2   A Viewer Site

FIGURE 4.3   A Reader Site

Using Zeldman's taxonomy, you can think about which group is most likely to come to your site. How can you meet their needs, while accomplishing your purpose?

■ TRY IT

Look back at the sites you located in the previous exercise. What kind of audience does the site seem designed for: users, viewers, or readers—or some new classification that we haven't yet included on the list?

WEB-WRITING STYLE

Once you've described your purpose and your probable audience, you can begin working on the hard part: writing your text. You can write the words that will appear on your pages at the same time you enter your HTML code. You can also write it separately if you want and then cut (or copy) and paste it into the pages you create. The advantage of this second writing approach is that it allows you to concentrate on your text without worrying, temporarily, about how to mark it up.

Page Length

One of your first considerations as you start writing your text will be how much text you need—how long should these pages be, anyway? There have always been different schools of thought on this question: are readers more likely to click or to scroll? However, Web usability studies have indicated that many readers, particularly casual ones, prefer short pages. Usability specialist Jakob Nielsen has pointed out that reading from a computer screen is around 25 percent slower than reading from a page; moreover, readers have reported that they feel uncomfortable reading long pages of text on screen. Nielsen argues that you should write around 50 percent less text for a computer page than for a comparable text page to account for both the slower reading speed and reader discomfort. Along with your shorter pages, you should also use shorter paragraphs. Again, readers need more accessible chunks of information on-screen so that they can assimilate the material quickly and move on.

However, this doesn't necessarily mean that you'll write 50 percent less text overall for your Web sites. It simply means that your text will be organized and presented differently on the Web than on a page. Rather than long, scrolling pages with multiple paragraphs, you'll probably create a series of shorter pages linked together so that a reader who wants to follow more deeply can choose a different path.

Think of ways to chunk your information into related segments. Then consider possible sequences in which those segments can be approached. (Chapters 7 and 8 discuss linking and site organization.)

The *Sun Microsystems Guide to Web Style* argues that any Web site that wants to grab a reader's attention (such as a sales site) must keep page length to the size of a single browser window (roughly 640 pixels by 480 pixels), so that a reader can see everything at a glance. Moreover, you'll need to check your page carefully in a browser to make sure that nothing extends beyond the edge of the window so that readers can get everything at once. Such pages can be presented as short chunks of information, linked clearly with navigation buttons or menus.

On the other hand, if you're constructing an informational site where you can assume a little more patience from potential readers, your pages can be longer. Scrolling allows a reader to maintain the context of your information better than following links; however, after a certain length (the Sun Web style guide argues the equivalent of four browser screens) even scrollers lose context and become frustrated.

It's a good idea to limit the length of even informational pages to no more than two and a half browser screens.

Anchors. If you write pages that are longer than a single screen (i.e., pages that require scrolling), it's essential to include anchor links on the page. Anchor links allow your readers to move up and down the page without scrolling through material, and they help to offset the problems of long pages. (See Chapter 7 for more on anchor links.) You should also provide a navigation bar (see Chapter 8) listing these anchor links so that your readers know what topics they can jump to.

▓ PRINTING WEB PAGES ▓

If you think your readers might want to print the text of your page (if, for example, you were providing the text of a report or a set of proposals), it's a courtesy to provide them with a printable version so that they don't have to print up everything you have on your screen (e.g., navigation buttons, banners, pictures, etc.). Copy the text from your page and save it in a separate HTML file. Then set up a page with white background and black text, including any necessary text formatting (e.g., headings, boldface, etc.) but without any graphics or other layout. Provide links to the printable version on each page of your report. It's particularly important to provide a print version of your pages if you're creating a page with a dark background and light type. If a reader tries to print such a page, the background color won't show up and the light-colored text may disappear against the light-colored paper!

You can also prepare PDF files for print versions of your text. PDF stands for Portable Document Format; it's a file format developed by Adobe Systems that allows you to send formatted documents over the Web. You must have access to Adobe Acrobat to prepare PDF files, and you should tell your readers to download Acrobat Reader in order to read them. (Acrobat Reader is a free program available for download at http://www.adobe.com/.)

Organization

If you follow these suggestions, your pages will be considerably shorter on the computer screen than they would be if you were printing them (although you'll probably have more pages than you would in an essay of comparable length). This length reduction should also be reflected in your organization: your Web writing should be as concise and clearly organized as possible.

A study by John Morkes and Jakob Nielsen confirms what you might already suspect: people read differently on the Web. Rather than reading every word, most people scan through pages, looking for cues such as headings and topic sentences in order to get the gist of the information. Rather than fighting against this tendency, you need to help your readers along by providing what they need to scan effectively.

Headings.   To make it easier for your readers to scan your pages, be sure to include at least two levels of headings: both a page heading to indicate the general topic of each page and subheadings to break up the page into supporting topics. Make sure that the headings themselves are informative: you want your reader to be able to grasp the general structure of your information by scanning the page. Finally, you should use the same level of headings for the same type of information throughout your site. That is, don't use a level two (<H2>) heading for a main point on one page and a level one (<H1>) heading for the same type of point on the next.

Topic Sentences. On paper, you can sometimes use climactic order for your paragraphs, saving your main point until the end. On the Web, *you must put your main point in the first or second sentence of each paragraph.* Frequently, people who scan read the first sentence of each paragraph, looking for the major point. If your major point is somewhere else, you may lose them.

Emphasis. If certain words or sentences are particularly important, consider highlighting them with color or even with boldfacing or italics. Again, your readers are scanning. Make sure they don't miss something that they need to know. However, don't go overboard: if too much is emphasized, nothing will stand out.

Inverted Pyramid Organization. According to Jakob Nielsen, the classic inverted pyramid style used by journalists is the most appropriate kind of organization for the Web. Inverted pyramid style works like this: begin with your conclusion, followed by the most important supporting information, then fill in the background details. If you use the inverted pyramid, even the reader who won't scroll (and some studies have indicated only 20 to 25 percent of your readers will) will at least get your major point before surfing on to the next page.

Concision. Most of all, consider the implications of writing 50 percent less text per page for the Web. How much text do you need to include on each page in order to convey your central point? Can you put your supplemental information into optional linked footnote pages that readers can choose to click, depending on their interest? Divide your information into sections: does it break down easily into separate pages? How much do you need on each page to help your reader to understand? Remember: on the Web you have only a few seconds to get your point across. How many words do you need per page to accomplish that purpose?

Feedback

Like any writing, Web pages benefit from feedback. Although you may receive reactions from your site visitors after you mount your pages, it's better to get as much feedback as you can before you go online. You can give paper copies of your text to other students to read and respond to, but it's best to let them see your pages as they'll appear on the Web. You want response not only to your words but also to your organization and design. Have them look at your pages through a browser and then ask for their reactions. These are some questions your readers can consider:

- Is it easy to find information on these pages, or are there problems in organization?
- Does the presentation match the subject? For example, if the pages discuss a serious topic, are the graphics and design appropriate?

- Is the text easy to read without strain?
- Is the subject clear, and is it presented with enough detail (whether on the individual pages or through links)?
- Is the organization of the site effective? Can you move around it easily? Can you find what you need?
- Did you understand the purpose the author had in creating this site?

Use your readers' reactions to revise your site until you're satisfied with the text, the design, and the site organization. Remember: all of those elements are part of writing for the Web.

Writing Style

Finally, we reach the bottom line in Web writing: the words themselves. How does (or should) your writing style on the screen differ from your writing style on the page? Drue Miller, a writer and designer with vivid studios, suggests several points to consider: tone, person, humor, and language technicalities.

Tone. The tone of the language on your site should match the tone of your site as a whole. If your topic and visual presentation are serious (e.g., a site for a nonprofit charity or a social–political appeal), your tone of voice should match. On the other hand, the Web is a much more personal medium than some print media. Many sites use an informal style. If you want to address your reader directly, make sure your tone of voice is light.

Person. Your tone will influence the decision on using *I* and *me* and *you* and *yours*. Many Web sites are written in first person, even using first person plural to represent an organization; the reader or user is addressed directly in second person. This use of first and second person is typical of informal business and technical writing, where the you attitude has long been recommended as a way of holding a reader's attention. Writing your information using personal pronouns may help to keep the reader involved with what you have to say. Moreover, the kind of passive voice, "subjectless" prose that's hard to read in most academic settings is even more excruciating on the Web. (For example, "Attention must be paid to the optimization of departmental resources through reduced photocopy costs.") Keep your prose active, direct, and, if possible, personal.

Humor. Catching the attention of a Web reader is a constant struggle, and sometimes humor can be an effective technique. Humor can help to explain complex topics, and it can make your site memorable. Unfortunately, it can also mark your site as dumb or even offensive. Miller's advice on the subject is sound: "Gentle pokes at yourself or your organization are highly effective means of conveying warmth and trust. Stay away from inside jokes, gratuitous puns, and tired old gags. And never make jokes at other people's expense.... Remember: There's a fine line between clever and stupid."

Language Technicalities. You should have a reference to rely on for questions of language and mechanics. *The Chicago Manual of Style* or *The Associated Press Stylebook and Libel Manual* both provide rules for questions of punctuation and capitalization. A good dictionary can help with spelling. A writing handbook, such as Christine A. Hult and Thomas N. Huckin's *New Century Handbook,* can answer questions about grammar.

■ REFERENCE BOOKS ON THE WEB

A variety of reference sources are available on the Web, but the question is whether it's easier to consult them or to consult a reference book. In general, if you have a dictionary or style book handy, it's a great deal easier to use than the Web, particularly if you're using a dial-up modem attached to a phone line. On the other hand, if you need something like a German dictionary and you have nothing like that on your bookshelf, then the Web is a ready alternative. Here are some basic Web references:

http://www.m-w.com/netdict.htm
An online version of the *Merriam Webster Collegiate Dictionary* that allows you to search for a term

http://www.facstaff.bucknell.edu/rbeard/diction.html
Web of Online Dictionaries: A multilingual dictionary site that allows you to look up words in several different languages through links to other online dictionaries

http://www.columbia.edu/acis/bartleby/bartlett/
Bartlett's Familiar Quotations

http://www.infoplease.com/
Information Please Fact Finder

http://www.m-w.com/thesaurus.htm
An online version of the *Merriam Webster Thesaurus*

http://www.ex.ac.uk/cimt/dictunit/dictunit.htm
Dictionary of Units: Listing of units of measurement with definitions and abbreviations

http://www.itools.com/research-it/
Research It: The ultimate look-up site, with dictionaries, thesauruses, geographical information, biographical information, and so on

http://www.dictionary.com/
A very user-friendly guide to dictionaries online.

http://www.cs.cmu.edu/references.html
A page of reference sources collected by the Carnegie Mellon Computer Science Department

Finally, *be sure to proofread your text before you post it!* In fact, let someone else proofread it too. It's difficult enough to catch errors when you're writing for print. Yet if you're writing text at the same time you're writing your HTML tags, it becomes even harder to catch mistakes.

Try printing out your text without HTML tags; give it to a friend or classmate to read it over, checking for both mechanical errors and clarity. Run the text (without HTML tags) through the spell check in your word processor. (Remember to save the text file as text only again when you're done.) You might even try reading the text aloud—it can help!

You may think that you've caught every mistake before you put your pages on the Web, but the one you overlooked will be the first thing you see after the pages are posted. Believe me, I know from experience!

WEB SITE ASSIGNMENT

Web Site 1 ■ A Text-Based Site

For your first Web site, your main content will be text rather than graphics, although you'll need to pay some attention to layout and design in order to make your pages readable and appealing. This will be what Jeffrey Veen, the author of *HotWired Style,* calls a "library site":

> The library has a vast collection of information available in numerous formats. It isn't concerned with the display of the information, just the method of organization. It expects that you know what kind of info you want, so its main goal is a clear organized system to help you find what you want as quickly as possible. It doesn't necessarily put the best books closest to the front door; it doesn't open their pages to give you a peek. (2)

You can do this site very simply, using the information on text tags and graphics tags provided in Chapters 2 and 3; however, you'll need to look at the information on linking provided in Chapter 7 to put your site together. And the advice on layout provided in Chapters 5 and 6 will also come in handy. Your teacher will decide how much she wants you to do.

TIPS

1. Pick a topic that you know something about, one about which you can write at least three Web pages. These sites will probably be mounted on your school's server, so the topic should be appropriate to the location (e.g., nothing libelous or obscene). It should also be a topic that might interest someone using the Web (e.g., not simply a paper you're writing for another class—unless the topic is one that would be of general interest).

2. Remember: a screen is not a page! People read differently online. Here are some suggestions:

- Edit your prose; keep your style tight and succinct (without descending to Dick and Jane level).
- Use active voice.
- Use first and second person if it's appropriate to the subject.
- Include lots of "navigation aids": titles, headings, subheads, lists. Make it easy for your readers to grasp your content quickly.
- Don't go too far with navigation aids, however. Don't fragment your prose so much that it's hard to reassemble your meaning.
- Get as much feedback as you can from other members of your class, as well as people outside your class. Ask others to read and respond to your text and to click though your site. Get their reactions both to what you're saying and the way your site is designed and organized.
- Proof everything! That includes text on buttons, on navigation links, even text that's imported as a graphic: mistakes can crop up anywhere!

3. Concentrate on organizing your information so that it isn't a burden to read. Remember that reading rates are slower on the Web. Consider ways of organizing and presenting the information so that your readers can access it quickly. Think about how you can help your readers navigate from page to page. (Do you want in-text links? A navigation bar? What should be linked to what?) Also consider the standard organization advice from the *Sun Microsystems Guide to Web Style:* even motivated readers have a hard time absorbing information on a computer screen. An effective page will be no longer than two and a half screens.

4. The best advice for coming up with an effective layout for a text-based site is to cruise through a few good ones. Keep these general guidelines in mind, however:
   - Lines of text that spread all the way across the screen are hard to read; follow the advice in Chapter 6 to create margins.
   - Your body type should be readable on any monitor; thus settings that reduce type below the default size (3) are a bad idea.
   - You want your text to be readable against your background; thus, don't use color combinations that make reading difficult or tiring (like red on blue, for example). Most experts still argue for dark text on a light background.

WEB SITES

Examples: User Sites

http://www.excite.com/
   Excite search engine

http:// www.nws.noaa.gov/
   The National Weather Service site

http://www.yahoo.com/
   Yahoo! search engine

Examples: Viewer Sites

http://www.graphic-home.iok.net/
   The Graphic Home

http://www.quokka.com/
   Quokkasports

http://www.studiomotiv.com/counterspace/
   Counterspace: A design site

http://www.T26font.com/
   T26 font design company

Examples: Reader Sites

http://www.feedmag.com/
   Feed magazine

http://www.legends.dm.net/
   Legends

http://www.salonmagazine.com/
   Salon magazine

http://www.word.com/
   Word magazine

Style Guides and Discussions Cited in This Chapter

http://home.netscape.com/computing/webbuilding/studio/feature19980827–
   1.html
   Drue Miller's "Click Here: How Not to Write for the Web"

http://home.netscape.com/computing/webbuilding/studio/feature19980827–
   4.html
   A roundtable discussion on Web writing from several writer–designers

http://www.alistapart.com/stories/who/
   Jeffrey Zeldman's "Design Your Audience"

http://www.useit.com/alertbox/9606.html
   Jakob Nielsen's "Inverted Pyramids in Cyberspace"

http://www.useit.com/alertbox/9703b.html
   Jakob Nielsen's "Be Succinct (Writing for the Web)"

http://www.useit.com/papers/webwriting/writing.html
Research report by John Morkes and Jakob Nielsen, "Concise, Scannable, and Objective: How to Write for the Web"

http://www.sun.com/styleguide/
Sun Microsystems Guide to Web Style

CHAPTER 5

LAYING IT ALL OUT
Tables

LAYOUT AND DESIGN

In this chapter we'll talk about structure—the structure of the information on your page. It's a mistake to think of layout, graphics, and color as decorations; they're part of the information you're conveying, and used effectively, they can enhance your reader's experience of your site. The layout of your pages will depend upon their content, of course; you'll use a different approach for a page that has a lot of text than for a page that includes several graphics. We'll talk more about the principles behind page design in Chapter 6; for now, we'll cover the nuts and bolts, the mechanics of setting up a table that can serve as a page grid.

The design of your pages requires a lot of advance planning and a lot of tagging. Unfortunately, HTML wasn't devised as a design medium; it was originally thought of simply as a way to convey text. Thus many of the techniques that designers have come up with to make Web page design more effective are essentially workarounds, ways of making data devices such as tables substitute for design grids. New developments, particularly the positioning capabilities of Cascading Style Sheets or CSS (see Chapter 9), may eventually replace these workarounds, but until CSS is fully implemented, we'll have to make do with what we have.

In this chapter we'll discuss how tables are set up in HTML and how you can make them function as containers for text and graphics.

WHY TABLES?

You may wonder why tables are necessary at all; can't you just use the HTML tags for text and graphics? You can, of course, but the results may not be what you really want. Let's say you followed the instructions in Chapter 3 and created a background stripe. Now you put it in your page and put some text on top of it. The result in a browser looks like Figure 5.1.

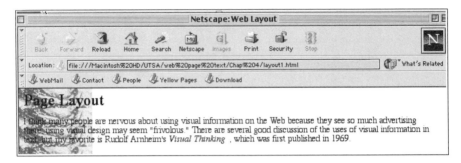

FIGURE 5.1   Text without Layout

Clearly, there are some problems here. The text on top of the textured stripe is hard to read, and it obscures the stripe pattern. You'd like to move the text out into the white part of the page, but how can you do that? To move the heading, you can use the **ALIGN** tag, but you can't use **ALIGN** with the text because you aren't aligning it with anything. Perhaps you could use **BLOCKQUOTE**, which would indent the text. If you aligned the heading and made the text a blockquote, Figure 5.2 shows how the result would look in a browser.

It's still not right, is it? The text is still on top of the stripe and now the heading looks off balance because the centered heading clashes with the asymmetric stripe. You could add another **BLOCKQUOTE** tag after the first one, nesting one blockquote inside another. That won't fix the heading, but it will indent the text a little more. However, it doesn't help much. The result in a browser looks like Figure 5.3.

You can go on adding blockquotes until the paragraph finally lines up on the white space. (That's how a lot of graphic-interface authoring programs indent text for you, which is another problem with these programs.) However, you won't really get the control you need over the text placement. Moreover, you won't be able to use the **BLOCKQUOTE** tags with a graphic because they

FIGURE 5.2   Text with Blockquote

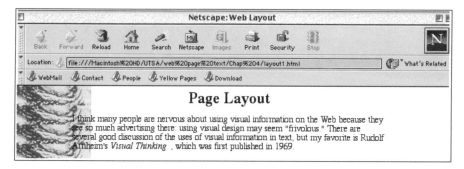

FIGURE 5.3   Text with Two Blockquotes

work only with text. Now Figure 5.4 shows the way things line up using a simple two-column, two-row table. (The table border is turned off by making it zero pixels wide, so you can't see the cells; we'll discuss borders further later in this chapter.) Lining the text up in this way is not that difficult to accomplish.

■ TRY IT

Go to some Web pages whose layout you like. Look at the HTML code. (In Internet Explorer, go to the View menu and select Source; in Netscape, go to View and select Page Source.) How many of these sites use tables for layout?

TABLES IN HTML

Before we get into using tables for layout, let's go over the basic table tags. HTML tables are actually a series of container tags, one set inside another. You

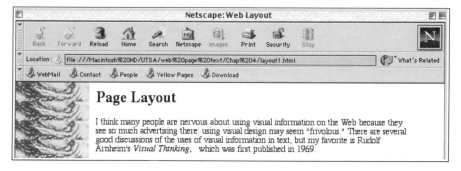

FIGURE 5.4   Text Layout with Table

begin and end a table with `<TABLE></TABLE>`. After this opening tag, you set up the rows, using `<TR></TR>`, which stands for Table Row. Within these rows, you can have two kinds of cells: Table Headers (`<TH></TH>`) and Table Data (`<TD></TD>`) cells. The only difference between these two cell types is that the text in header cells is boldfaced and centered in the cell, whereas the text in data cells is plain and left justified. The HTML for a simple three-column, two-row table looks like this. (The table border is a little thicker than the default setting, so you can see it more clearly.)

```
<HTML>
<HEAD>
        <TITLE>Table 1</TITLE>
</HEAD>
<BODY>
<TABLE BORDER=3>
<TR>
<TH>First cell (header)</TH> <TD>Second
cell</TD><TD>Third cell</TD></TR>
<TR>
<TH>Second header cell</TH> <TD>Another data
cell</TD><TD>Last data cell</TD></TR>
</TABLE>
</BODY>
</HTML>
```

Figure 5.5 shows what it looks like in a browser.

Creating a Table

1. Type `<TABLE>` (If you want to see the cell divisions, type `<TABLE BORDER="2">`)
2. Type `<TR>` to define the beginning of the first row.

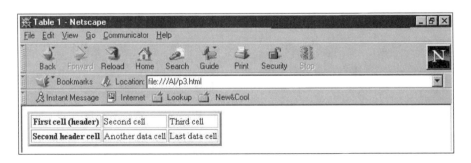

FIGURE 5.5   Simple Table

3. Type <TH> to create a header cell in the first column *or* <TD> to create a data cell.
4. Type the contents of the cell and </TH> *or* </TD>
5. Repeat Steps 3 and 4 for each cell in the row.
6. Complete the first row by typing </TR>
7. Repeat Steps 2 through 6 for each additional row.
8. Finish the table by typing </TABLE>

■ TRY IT

Create the HTML and text for a page with a simple table. Include at least two rows <TR>, two header cells <TH>, and two data cells <TD>

Remember to include the basic template: <HTML></HTML>, <HEAD></HEAD>, <TITLE></TITLE>, <BODY></BODY>

If you're working on the text-based site assignment, you can use these exercises to try out table effects with the text and graphics for your site.

Table Spacing

You can add two different types of spacing to your tables, depending on where you want the space to be placed. *Cell padding* adds space around the information inside the cell; *cell spacing* adds space between cells (i.e., outside the cell). Cell spacing makes the table bigger without enlarging the individual cells.

Both measurements are added to the TABLE tag.

Changing the Table Spacing

1. To add cell *padding* around the contents of a cell, after <TABLE, type CELLPADDING= and a number for the number of pixels of space you want. (The default cell padding is 1.)
2. To add cell *spacing* between cells, after <TABLE, type CELLSPACING= and a number for the number of pixels of space you want. (The default cell spacing is 2.)

■ TRY IT

Use the table you created in the previous exercise. First add some cell padding, then add some cell spacing. Be sure to check the result in a browser each time.

Table Dimensions

You can use the WIDTH and HEIGHT attributes to set the width and/or height of your entire table or of cells within the table, which allows you to set the size for individual cells or the table as a whole. Like padding and spacing, the dimensions of the entire table can be defined in the TABLE tag.

You can change the dimensions of individual cells to make them uneven widths or heights. (If you don't specify width or height, the cells will all have the same size; however, the largest cell in the row or column will influence the width or height of the other cells, something we'll discuss in more detail later in this chapter.)

Changing Dimensions

1. To resize the *whole table,* after <TABLE, type WIDTH=X HEIGHT=Y>, adding numbers of pixels for the height and width. You can also use percentages that indicate how big the table should be with respect to the full window.
2. To resize an *individual cell,* place the cursor inside the cell tag (i.e., either TH or TD), and type WIDTH=x HEIGHT=y, adding numbers of pixels for the height and width. You can also use percentages to indicate how big the cell should be with respect to the full table size.

Now let's add these tags to the basic HTML we had above. First make the whole table wider by making it 500 pixels wide, and then divide those 500 pixels into three columns of 200, 150, and 150 pixels. The HTML looks like this:

```
<HTML>
<HEAD>
        <TITLE>Table 1</TITLE>
</HEAD>
<BODY>
<TABLE BORDER="3" WIDTH="500" BGCOLOR="#FFFFFF">
<TR>
<TH WIDTH="200">First cell (header)</TH> <TD
WIDTH="150">Second cell</TD><TD WIDTH="150">
Third cell</TD></TR>
<TR>
<TH>Second header cell</TH> <TD>Another data
cell</TD><TD>Last data cell</TD></TR>
</TABLE>
</BODY>
</HTML>
```

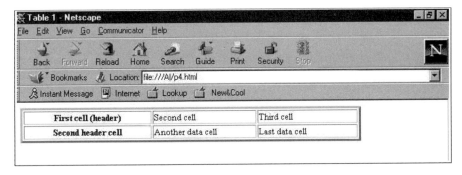

FIGURE 5.6 Changed Dimensions

Notice that the widths added in the first row also apply in the second row, or, to put it another way, the widest cell sets the column width. In a browser it looks like Figure 5.6. (A white background color is added so that you can see the cells a little more clearly.)

Now we'll add some cell padding—that is, space inside each cell. The HTML looks like this:

```
<HTML>
<HEAD>
        <TITLE>Table 1</TITLE>
</HEAD>
<BODY>
<TABLE BORDER="3" WIDTH="500" BGCOLOR="#FFFFFF"
CELLPADDING="5">
<TR>
<TH WIDTH="200">First cell (header)</TH> <TD
WIDTH="150">Second cell</TD><TD WIDTH="150">
Third cell</TD></TR>
<TR>
<TH>Second header cell</TH> <TD>Another data
cell</TD><TD>Last data cell</TD></TR>
</TABLE>
</BODY>
</HTML>
```

Figure 5.7 shows what it looks like in a browser.

Now we'll change the cell padding to cell spacing—that is, space around each cell. The HTML would be:

```
<HTML>
<HEAD>
```

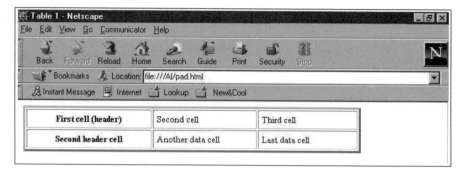

FIGURE 5.7   Cell Padding

```
        <TITLE>Table 1</TITLE>
</HEAD>
<BODY>
<TABLE BORDER="3" WIDTH="500" BGCOLOR="#FFFFFF"
CELLSPACING="10">
<TR>
<TH WIDTH="200">First cell (header)</TH> <TD
WIDTH="150">Second cell</TD><TD
WIDTH="150">Third cell</TD></TR>
<TR>
<TH>Second header cell</TH> <TD>Another data
cell</TD><TD>Last data cell</TD></TR>
</TABLE>
</BODY>
</HTML>
```

In the browser it looks like Figure 5.8.

You may not be able to see the effects of cell spacing as easily as the effects of cell padding. It makes the divisions between the cells thicker.

Aligning Cell Contents

You can change the alignment of cell contents (either a header or a data cell) both horizontally and vertically. Header cells are normally centered and aligned to the cell top; data cells are aligned to the top left. The two attributes involved are ALIGN (for horizontal alignment) and VALIGN (for vertical alignment).

Changing Horizontal Alignment

1. Place the cursor inside the initial tag for each cell (i.e., <TH or <TD).

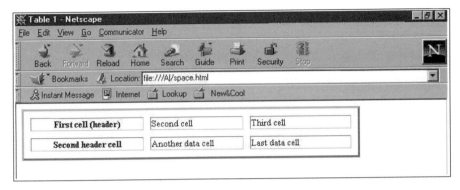

FIGURE 5.8   Cell Spacing

2. Type ALIGN=LEFT, RIGHT, *or* CENTER> *or*
3. Align the entire row by placing your cursor inside the <TR tag and typing ALIGN=LEFT, RIGHT, *or* CENTER>

Note: You may need to increase the size of the table to see the horizontal alignment.

Changing Vertical Alignment
1. Place the cursor inside the initial tag for each cell (i.e., <TH or <TD).
2. Type VALIGN=TOP, MIDDLE, *or* BOTTOM> *or*
3. Align the entire row by placing your cursor inside the <TR tag and typing VALIGN=TOP, MIDDLE, *or* BOTTOM>

Note: You may need to increase the size of the table to see the vertical alignment.

■ TRY IT

Write the HTML and text for a page with a table; this time, define the table's height and width. Define the height and/or width of at least two cells as well (be careful not to exceed the height or width you specified for the whole table). Align the contents of a cell or of an entire row either horizontally or vertically. Remember to include the basic template: <HTML></HTML>, <HEAD></HEAD>, <TITLE></TITLE>, <BODY></BODY>

Other Table Effects: Borders, Captions, and Background Color

The border on a table in HTML is represented by shaded lines in the browser, giving a kind of three-dimensional effect. You can change the width of the border by changing the number of pixels, using BORDER as an attribute within the TABLE tag and giving the number of pixels as a value. The default for table borders is none; that is, the borders shouldn't show up in the browser unless you add a value for the table border. However, when you're using a table for layout purposes, it's a good idea to make sure the borders won't show by using BORDER=0 in the TABLE tag.

Changing the Table Border

1. After <TABLE, type BORDER
2. If you want to specify thickness in pixels, add = and a number (e.g., BORDER=2). The result should look like this: <TABLE BORDER=2>
3. If you want to make sure you have *no* visible border on your table, type BORDER=0

Sometimes you may want to give your table a title. You can do this by adding a CAPTION container tag after the TABLE tag. The caption will appear above the table; if you want it to appear below the table instead, use ALIGN as an attribute within the CAPTION tag and give the value as "bottom".

Creating a Table Caption

1. After <TABLE> but before any row or cell tags, type <CAPTION>
2. Type the caption for the table.
3. Type </CAPTION>

Note: If you want the caption to appear below the table rather than above it, type <CAPTION ALIGN="bottom">

Tables can have a background that's a different color from the rest of the page, which can create some interesting effects; you can also add different colors to different cells. Just keep in mind that a changed background on a table will show the cell borders even if they've been turned off. (The background doesn't extend into the border areas.) How to get rid of that border is covered later in this chapter.

Changing the Table Background

1. *To color an individual cell,* place the cursor inside the initial tag (i.e., <TH or <TD) and type BGCOLOR= either a hexadecimal color designation or one of the sixteen predefined colors.

2. *To color the entire row,* place the cursor inside the <TR tag and type BG-COLOR= either a hexadecimal color designation or one of the sixteen pre-defined colors.

3. *To color an entire table,* place the cursor inside the <TABLE tag and type BGCOLOR= either a hexadecimal color designation or one of the sixteen predefined colors.

Font size and color are set within the cells of the table. Any text formatting that you entered earlier—for example, in the BODY tag—will have no effect on table text. So if you need to enlarge the text size or change the text color, do it by using a FONT tag after the TD or TH tag.

Changing the Font Size or Color on a Table

1. Set up your table, using <TABLE><TR>
2. Type *either* <TD> *or* <TH>
3. Type <FONT
4. To change font size, type SIZE="n"> (a number from 1 to 7)
5. To change font color, type COLOR= a hexadecimal color designation or one of the sixteen predefined color names>
6. Repeat for each cell where you want to change the type size or color.

Here's what the HTML would look like for a table using these tags. (The page background is changed to make the cells easier to see.)

```
<HTML>
<HEAD>
        <TITLE>Table 1</TITLE>
</HEAD>
<BODY BGCOLOR="#FFFFFF">
<TABLE BORDER="0" WIDTH="500" CELLSPACING="10">
<CAPTION>Table With Altered Text and
Background</CAPTION>
<TR BGCOLOR="#33FFFF">
<TH WIDTH="200"><FONT SIZE="5">First cell
(header)</FONT></TH> <TD WIDTH="150"> <FONT
SIZE="5">Second cell</FONT></TD><TD
WIDTH="150">Third cell</TD></TR>
<TR>
<TH BGCOLOR="#33FFFF">Second header cell</TH>
<TD BGCOLOR="#FFFF00">Another data cell</TD><TD
BGCOLOR="#FFFF00">Last data cell</TD></TR>
</TABLE>
</BODY>
</HTML>
```

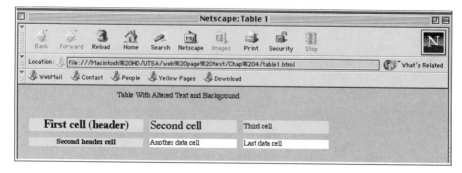

FIGURE 5.9   Table with Captions and Background Color

In a browser, it would look like Figure 5.9. (Notice that even though the border is set to zero, the colored backgrounds of the cells show where the border is.)

You can see the full-color version at the book Web site, at http://www. abacon.com/batschelet/.

■ TRY IT

Use the table you created in the previous exercise or write the HTML for a new table. Change the border size, the font size and/or color, and the background color for the table. Include a table caption. Remember to include the basic template: <HTML></HTML>, <HEAD></HEAD>, <TITLE></TITLE>, <BODY></BODY>

TABLES FOR LAYOUT

HTML tables allow you to create fairly complex page grids, giving you more control over the placement of your graphics and your text. By using spacers to separate elements, you can also use tables to create empty white space.

It's a good idea to sketch a rough draft of the table page grid you want to create before you start writing the code. That way you can work out the relationship you want between the elements on the page. Because the tags you write don't really look like a table, it will help to have a clear mental image of what you want the page to look like and how the columns will line up—particularly when you start using tags such as **COLSPAN** and **ROWSPAN**. You can try this with the page grid for your text-based site; rough out a table grid as you go through the rest of the chapter.

Creating Columns

Using tables to create columns of text is fairly simple, and it's a good way to get a sense of what table layout is all about. Let's start with a three-column page.

The table background is set to white and the borders are turned off. This is the HTML:

```
<HTML>
<HEAD>
        <TITLE>Columns</TITLE>
</HEAD>
<BODY>
<TABLE BORDER="0" BGCOLOR="#FFFFFF">
        <TR>
                <TD> Some Web pages are mostly
text. This isn't necessarily bad, but it can
create problems if you end up with long lines of
type. Nobody likes scrolling. </TD>
                <TD> Columns are one way around the
plain text problem. One thing to keep in mind
(especially if you're accustomed to page layout
programs): you can't "flow" text from one column
to the next. Thus you need to be aware of the
relative size of each column.</TD>
                <TD> Columns don't really allow
more words onto the page, but they allow more
white space.</TD>
        </TR>
</TABLE>
</BODY>
</HTML>
```

Figure 5.10 shows what it looks like in a browser.

Because no dimensions are specified for the table, notice how it has stretched to fill the browser window here. HTML table cells are elastic: they can stretch to any size. Because the browser window is set to almost full screen, the three columns have divided in a way that gives them equal vertical space, although the middle column (which has more text) has stretched to fill

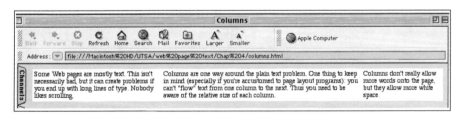

FIGURE 5.10 Columns

up more horizontal space. Figure 5.11 shows what happens if the browser is set to a narrower window size (as with a smaller or lower-resolution monitor).

Now the columns have unequal widths and don't line up at the top. The default vertical alignment for columns is middle; thus unless you indicate the vertical alignment you want, the tops of your columns may be uneven. Now let's see what happens when we make each cell of the table the same size by entering a `WIDTH` value; each column aligns at the top, using `VALIGN`. The HTML looks like this:

```
<HTML>
<HEAD>
        <TITLE>Columns</TITLE>
</HEAD>
<BODY>
<TABLE BORDER="0" BGCOLOR="#FFFFFF" WIDTH="630">
        <TR VALIGN="top">
                <TD WIDTH="210"> Some Web pages are
mostly text. This isn't necessarily bad,
but it can create problems if you end up with
long lines of type. Nobody likes scrolling. </TD>
                <TD WIDTH="210"> Columns are one
way around the plain text problem. One thing to
keep in mind (especially if you're accustomed to
page layout programs): you can't "flow" text from
one column to the next. Thus you need to be aware
of the relative size of each column.</TD>
                <TD WIDTH="210"> Columns don't
really allow more words onto the page, but they
allow more white space.</TD>
        </TR>
</TABLE>
```

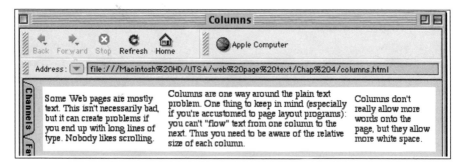

FIGURE 5.11 Narrower Window

```
        </BODY>
        </HTML>
```

In the browser, it looks like Figure 5.12.

Now let's add something else to the columns: column heads. We'll do this by adding another row (with the same cell widths) and using header cells. The HTML looks like this:

```
<HTML>
<HEAD>
        <TITLE>Columns</TITLE>
</HEAD>
<BODY>
<TABLE BORDER="0" BGCOLOR="#FFFFFF" WIDTH="630">
        <TR>
                        <TH WIDTH="210">
Text Pages
                </TH>
                        <TH WIDTH="210">
Using Columns
                </TH>
                        <TH WIDTH="210">
No Flow
                </TH>
        </TR>
        <TR VALIGN="top">
                        <TD WIDTH="210"> Some Web pages are
mostly text. This isn't necessarily bad, but it
can create problems if you end up with long lines
of type. Nobody likes scrolling. </TD>
                        <TD WIDTH="210"> Columns are one
way around the plain text problem. One thing to
keep in mind (especially if you're accustomed to
```

FIGURE 5.12 Columns with VALIGN

```
page layout programs): you can't "flow" text from
one column to the next. Thus you need to be aware
of the relative size of each column.</TD>
                <TD WIDTH="210"> Columns don't
really allow more words onto the page, but they
allow more white space.</TD>
        </TR>
</TABLE>
</BODY>
</HTML>
```

Figure 5.13 shows what it looks like in the browser.

If desired, you can use different colors for the background of those cells, so that the heads are one color and the columns another. Adding the BGCOLOR attribute to the TR tag for each row makes the page in the browser look like Figure 5.14.

Now let's get rid of those lines between the columns (which show up even though the border is set to zero). To eliminate the lines between columns and

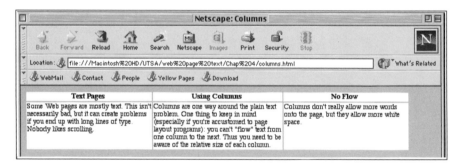

FIGURE 5.13   Column Heads

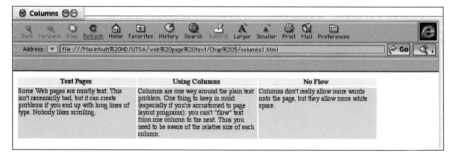

FIGURE 5.14   Column Colors

rows, we'll set the CELLSPACING to zero. The HTML looks like this. (Again, the page background is changed to white to make the cells easier to see.)

```
<HTML>
<HEAD>
        <TITLE>Columns</TITLE>
</HEAD>
<BODY BGCOLOR="#FFFFFF">
<TABLE BORDER="0" WIDTH="630" CELLSPACING="0">
        <TR BGCOLOR="#FFFF33"><TH WIDTH="210">
Text Pages
                </TH>
                        <TH WIDTH="210">
Using Columns
                </TH>
                        <TH WIDTH="210">
No Flow
                </TH>
        </TR>
        <TR VALIGN="top" BGCOLOR="#00FFFF">
                <TD WIDTH="210"> Some Web pages are
mostly text. This isn't necessarily bad, but it
can create problems if you end up with long lines
of type. Nobody likes scrolling. </TD>
                <TD WIDTH="210"> Columns are one
way around the plain text problem. One thing to
keep in mind (especially if you're accustomed to
page layout programs): you can't "flow" text from
one column to the next. Thus you need to be aware
of the relative size of each column.</TD>
                <TD WIDTH="210"> Columns don't
really allow more words onto the page, but they
allow more white space.</TD>
        </TR>
</TABLE>
</BODY>
</HTML>
```

In a browser it looks like Figure 5.15.

There's one more problem to deal with here. There's not enough space between the columns now—they almost seem to run into each other. To take care of that, we add some CELLPADDING, which will add space between columns without adding lines. You'll put the CELLPADDING attribute in the TABLE tag, along with the CELLSPACING=0, and set the value for ten pixels.

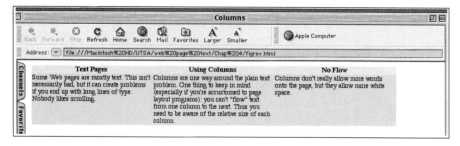

FIGURE 5.15   No Cell Spacing

Figure 5.16 shows how the result looks in a browser (see the full-color versions at the book Web site found at http://www.abacon.com/batschelet/).

To jazz this up some more, you could add more text to columns one and three to try to make the three columns line up more evenly, or you could make the cells different colors. However, this gives you the basic idea behind using tables for layout: they allow you more control over where information appears on your pages.

■ TRY IT

Write the HTML and text for a simple three-column layout table. Be sure to set the borders and cell spacing to zero. Add other layout features, such as headers and colored backgrounds, if you wish.

Remember to include the basic template: <HTML></HTML>, <HEAD></HEAD>, <TITLE></TITLE>, <BODY></BODY>

If columns are appropriate for your Web site, incorporate them into your grid.

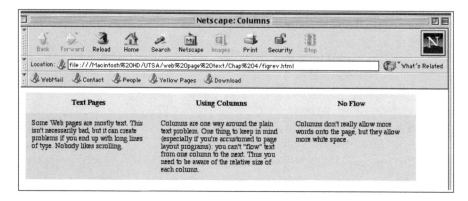

FIGURE 5.16   Cell Padding

Creating White Space

You can use table cells to create margins, both at the top and on the left of your text (although CSS provides a simpler way of doing this, which we'll cover in Chapter 9). You do this by creating empty table cells that create space around other table cells. In reality, of course, these cells aren't empty: if they were they'd simply collapse. Although there are several ways to fill the cells, one of the simplest is to use a *nonbreaking space,* a special HTML character. Nonbreaking spaces—which are written as —were originally meant to keep elements together so that they wouldn't be separated at the ends of lines. Say, for example, you wanted to keep your first and last names—Albert Einstein—together so that they didn't separate at the end of a line. In HTML you could code your name this way: Albert Einstein.

Actually, there are many more of these so-called special characters that you can use to do things; for example, putting accents in words such as *résumé* (see the Special Characters box). To create a column that would be blank but that wouldn't collapse, you'd write these tags:

```
<TABLE WIDTH="640" BORDER=0>
<TR>
<TD WIDTH="150"> </TD>
```

■ SPECIAL CHARACTERS ■

There are several special characters you can use to insert symbols such as © or ® or diacritical marks such as accents. All of these characters begin with & and end with ;. Here are some of the more common ones:

```
é  =  &eacute;
á  =  &aacute;
è  =  &egrave;
à  =  &agrave;
ü  =  &uuml;
ñ  =  &ntilde;
ø  =  &oslash;
©  =  &copy;
®  =  &reg;
```

In Figure 5.4 earlier in this chapter, I used a nonbreaking space to create white space over the patterned stripe in my background. I entered a WIDTH of 150 for the cell containing the nonbreaking space because the stripe is about 100 pixels wide; thus I had 50 extra pixels of space between the stripe and the text. I added another cell for the heading, and then a second row containing another nonbreaking spacer and the text. The HTML looks like this:

```
<HTML>
<HEAD>
       <TITLE>Web Layout</TITLE>
</HEAD>
<BODY BACKGROUND="stripe.gif">
<TABLE WIDTH="640" BORDER="0">
<TR>
<TD WIDTH="150"> </TD>
<TD><H1>Page Layout</H1></TD></TR><TR>
<TD WIDTH="150"> </TD>
<TD><FONT SIZE="4">I think many people are
nervous about using visual information on the Web
because they see so much advertising there: using
visual design may seem "frivolous." There are
several good discussions of the uses of visual
information in text, but my favorite is Rudolf
Arnheim's <EM>Visual Thinking,</EM> which was
first published in 1969.</FONT></TD></TR>
</TABLE>
</BODY>
</HTML>
```

Setting the border to one pixel allows you to see the space that has been filled with the nonbreaking spaces. Figure 5.17 shows how it looks in the browser.

■ TRY IT

Write the HTML and text for a page that uses a table with nonbreaking spaces to create a left margin. Remember to set the table border to zero and to include the basic template: <HTML></HTML>, <HEAD></HEAD>, <TITLE></TITLE>, <BODY></BODY>

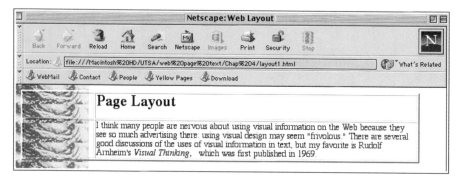

FIGURE 5.17 Using Spacers

If you need white space for your page grid, try using empty table cells using nonbreaking spaces.

Laying Out Text and Graphics

You can use tables to line up text and graphics side-by-side, so that you don't have to deal with text wrap or alignment. A simple layout with a colored text block and graphic would look like this:

```
<HTML>
<HEAD>
        <TITLE>Berries</TITLE>
</HEAD>
<BODY BGCOLOR="#FFFF99">
        <TABLE BORDER="0" CELLSPACING=0
CELLPADDING=10 WIDTH="350" ALIGN=Center>

        <TR>
                <TD WIDTH="250" BGCOLOR="#FF9900"
VALIGN="top"> <FONT SIZE="4">
It's important to keep track of the size (in
pixels) of your graphics. When you place them in
table cells you'll want the cells to be the right
size to hold them. You also need to make sure
that the total width of all the columns in your
table doesn't exceed the total width of the table
(if you've entered one).
</FONT></TD>
                <TD WIDTH="100" BGCOLOR="#FFFFFF">
<IMG SRC="berries.gif"></TD>
        </TR>
</TABLE>
</BODY>
</HTML>
```

In a browser, you'd see the simple graphic layout of Figure 5.18.

■ TRY IT

Write the HTML and text for a page that uses a table to align a graphic with a block of text. Remember to set the table border to zero and to include the basic template: <HTML></HTML>, <HEAD></HEAD>, <TITLE></TITLE>, <BODY></BODY>

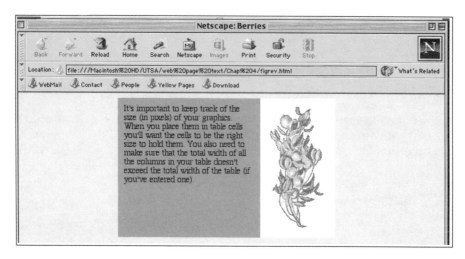

FIGURE 5.18   Simple Graphic Layout

If you need to align text and graphics for your page grid, try using these tags.

COLSPAN and ROWSPAN

Layouts with multiple text and graphic cells can become more complex. As you saw above, the tallest cell determines the height of the cells in a row, and the tallest element determines the tallest cell. If you have a large graphic in one cell and smaller amounts of text in others (or multiple graphics), the size of the large graphic will determine the size of the other cells (and may give you extra space you don't want in the row). To make taller and shorter cells more independent of each other, you can use the COLSPAN and ROWSPAN attributes.

For example, a tall graphic next to another graphic (in this case, a heading created in a graphics program) and a text passage ends up looking like Figure 5.19.

I really want the graphic to line up with both the heading and the text, with the heading at the top of the graphic rather than at the middle. In other words, I want the tall graphic to span two rows, the row with the heading and the row with the text. I want the table cells to line up like this:

FIGURE 5.19   A Big Graphic

To accomplish that, I'll need to use the **ROWSPAN** tag. **COLSPAN** and **ROWSPAN** allow one table cell to spread across several columns or across several rows. You can use them in your layout tables to place wide or long graphics, or to create interesting variations in line length or color blocks. In this example with a tall graphic along with a smaller graphic and a text block, you can place the tall graphic in the first cell and use the rowspan command to stretch that cell across two rows, so that it doesn't stretch the cells next to it. **COLSPAN** and **ROWSPAN** are attributes inserted into the **TD** or **TH** tags; you also include a value for the number of columns or rows you want the cell to stretch across. In the case of the big graphic in Figure 5.19, the tag would read like this:

```
<TD ROWSPAN="2"><IMG SRC="cookie.gif"></TD>
```

Inserting this tag into HTML results in the following. (I'm also changing the vertical alignment so that the text is at the top of its cell and the heading is at the bottom of its cell.)

```
<HTML>
<HEAD>
        <TITLE>Big Graphic</TITLE>
```

```
</HEAD>
<BODY BGCOLOR="#FFFF00">
<TABLE BORDER="0" WIDTH="640" CELLSPACING="0">
        <TR>
                <TD ROWSPAN="2"><IMG SRC=
"cookie.gif"></TD>
                <TD VALIGN="bottom"><IMG SRC=
"Finding.gif"></TD></TR>
                <TR>
                <TD VALIGN="top"><FONT SIZE="4"
COLOR="#006633">Finding graphics for the Web is
sometimes challenging, but a lot of good-looking
drawings are now out of copyright, and thus can
be used for clipart. This drawing is from a
collection of clipart from the Victorian era. It
was probably a magazine illustration originally.
I love the man's expression as he eats his
cookie!</FONT></TD>
        </TR>
</TABLE>
</BODY>
</HTML>
```

In the browser it looks like Figure 5.20.

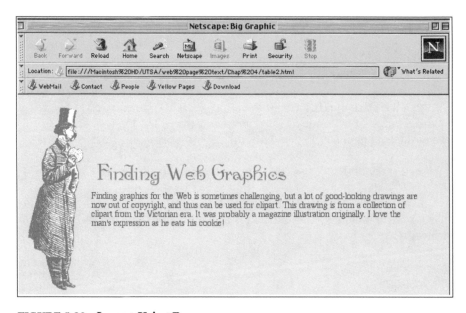

FIGURE 5.20 Layout Using Rowspan

Although the result is closer to what I want, it's still too "sprawly": I want the text block to be more compact and I still want that heading to move up more. So I'm going to specify a HEIGHT for the top row of the table (to move the heading up by making the row narrower) and a WIDTH for the table as a whole. Finally, I'm going to add another row at the bottom of the table with a spacer to span the large graphic over three rows rather than two. The table layout will now look like this:

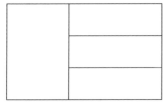

The HTML will now look like this:

```
<HTML>
<HEAD>
      <TITLE>Big Graphic</TITLE>
</HEAD>
<BODY BGCOLOR="#FFFF00">
<TABLE BORDER="0" WIDTH="500" CELLSPACING="0">
      <TR>
            <TD ROWSPAN="3"><IMG SRC=
"cookie.gif"></TD>
            <TD VALIGN="Bottom" HEIGHT="40">
<IMG SRC="Finding.gif"></TD></TR>
            <TR>
            <TD VALIGN="Top" WIDTH="350" HEIGHT
="70"><FONT SIZE="4" COLOR="#006633">Finding
graphics for the Web is sometimes challenging,
but a lot of good-looking drawings are now out of
copyright, and thus can be used for clipart. This
drawing is from a collection of clipart from the
Victorian era. It was probably a magazine
illustration originally. I love the man's
expression as he eats his cookie!</FONT></TD>
      </TR> <TR>
            <TD WIDTH="350">
 
            </TD>
      </TR>
</TABLE>
</BODY>
</HTML>
```

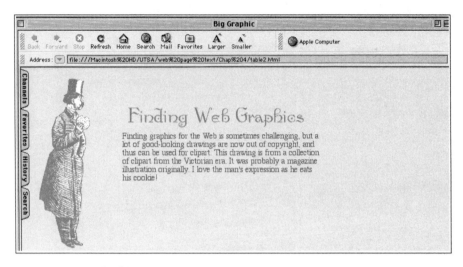

FIGURE 5.21   Final Layout

In a browser the result is shown in Figure 5.21.

There's one tricky thing to remember about COLSPAN and ROWSPAN—you *must* reduce the number of columns or rows to make up for the columns or rows that the COLSPAN or ROWSPAN is stretching across. In the example above, we started with two columns and two rows (a spacer cell was to the left of the text originally). In converting the large graphic to span two rows, the first cell in the second row (which was now occupied by the graphic) was subtracted. Before, the table looked like this, with four cells:

Afterward, with the graphic taking up two cells, it looks like this:

Because there's one fewer row cell in the table, the HTML code will have to reflect that fact with one fewer cell in the tags.

Instead of

```
<TABLE>
        <TR>
                <TD> Row_1_Cell_1_ </TD>
```

```
                    <TD> Row_1_Cell_2_ </TD>
          </TR>
          <TR>
                    <TD> Row_2_Cell_1_ </TD>
                    <TD> Row_2_Cell_2_ </TD>
          </TR>
</TABLE>
```

the tags would be

```
<TABLE>
          <TR>
                    <TD ROWSPAN="2"> Row_1_Cell_1_ </TD>
                    <TD> Row_1_Cell_2_ </TD>
          </TR>
          <TR>
                    <TD> Row_2_Cell_1_ </TD>
          </TR>
</TABLE>
```

Whenever you use COLSPAN or ROWSPAN, remember to reduce the number of columns or rows by the number of cells that are being taken up by the stretched cell. For example, if you have a colspan of three, you'll have two fewer cells in the row; if you have a rowspan of two, you'll have one fewer cell in the second row.

Using COLSPAN

1. Type either <TH or <TD depending on whether you want the cell to be a header or a data cell.
2. Type COLSPAN= the number of columns the cell will spread across>
3. Type the cell's contents and either </TH> or </TD>
4. Complete the table; remember to subtract the number of columns that the expanded cell spans across from the total number of cells in the row. (For example, if there were five cells in the row and the first cell stretches across two, you'll do tags for only three more cells, rather than four.)

Using ROWSPAN

1. Put the cursor inside the cell that will span more than one row.
2. Type <TH or <TD depending on the type of cell you want.
3. Type ROWSPAN= the number of rows the cell will spread across>
4. Type the cell's contents and </TH> or </TD>
5. Complete the table; remember to subtract the number of cells that the expanded cell spans across from the total number of cells. (For example, if

there were five rows and the first cell stretches across two, you'll do tags for only three rows more, rather than four.)

▪ TRY IT

Write the HTML and text for a table that includes either `COLSPAN` or `ROWSPAN`. Remember to set the table border to zero and to include the basic template: `<HTML></HTML>`, `<HEAD></HEAD>`, `<TITLE></TITLE>`, `<BODY></BODY>`

If `COLSPAN` or `ROWSPAN` would help you to set up your page, add it to your grid.

This chapter introduces you to the use of tables for layout. You can gain more experience by experimenting with layout yourself. Keep looking at the HTML code on pages with attractive layout; notice how the designers use tables. Then try adapting some of those designs to your own pages. See how you could use tables to create the kind of design you're looking for. Try sketching out your page layout in table cells first (using pencil and paper—the old-fashioned way!) and then setting up the cells in HTML. At first, the process may seem clumsy, but after you've worked with tables awhile, they'll seem less cumbersome. They'll also help you put together pages you can be proud of. Once you have a sense of the way tables work for layout, you can use an authoring program to do the tag-intensive part of the process, creating the tables for you. But you can still work with the table code yourself to set it up the way you want it. That's the advantage of knowing HTML.

▪ WRITING ASSIGNMENT

Design a Web page using a table for layout. Your table should have at least *three* rows and at least *two* columns. It should include at least *one* graphic (your instructor may specify more) and at least *three* paragraphs of text. Your table should also

- Have no visible borders
- Use one or more nonbreaking spaces for spacers

Tips   If you're working on the text-based site assignment, you can use this assignment to create a sample page for it—an illustrated definition, for example, or an explanation of a process or procedure. You can also use the information in this chapter to set up an attractive, readable layout for your text-based site: you can use a table to create a left margin, so that your text line isn't too wide to read easily.

TAGS USED IN CHAPTER 5

OPENING TAG	CLOSING TAG	EFFECT
`<TABLE>`	`</TABLE>`	Marks the beginning and end of the table
`<TR>`	`</TR>`	Marks the beginning and end of the table row
`<TH>`	`</TH>`	Creates a header cell
`<TD>`	`</TD>`	Creates a data cell
`CELLSPACING=n*`	None	Puts a number of pixels of space between table cells (attribute in **TABLE** tag)
`CELLPADDING=n*`	None	Puts a number of pixels of space around the contents of a table cell (attribute in **TABLE** tag)
`WIDTH=n*`	None	Sets the width for an entire table or for an individual cell (attribute in **TABLE** *or* **TH** *or* **TD** tag)
`HEIGHT=n*`	None	Sets the height for an entire table or for an individual cell (attribute in **TABLE** *or* **TH** *or* **TD** tag)
`ALIGN=left, right,` *or* `center`	None	Sets horizontal alignment for entire row or for an individual cell (attribute in **TABLE** *or* **TR** *or* **TH** *or* **TD** tag)
`VALIGN=top, middle,` *or* `bottom`	None	Sets vertical alignment for entire row or for an individual cell (attribute in **TABLE** *or* **TR** *or* **TH** *or* **TD** tag)
`BORDER=n*`	None	Sets width of border around table and between cells (for no border set **BORDER=0**; attribute in **TABLE** tag)
`<CAPTION>`	`</CAPTION>`	Creates table title (aligned at top; for bottom alignment use **ALIGN=bottom**)
`BGCOLOR=x**`	None	Inserts a background color (attribute in **TABLE** *or* **TR** *or* **TH** *or* **TD** tag)
` `	None	Inserts a nonbreaking space (used for a spacer in an empty table cell)
`COLSPAN=n***`	None	Stretches a cell across more than one column (attribute in **TH** *or* **TD** tag)
`ROWSPAN=n***`	None	Stretches a cell across more than one row (attribute in **TH** *or* **TD** tag)

*Insert a number of pixels.
**Insert a hexadecimal color number or one of the sixteen predefined colors.
***Insert the number of rows or columns the cell will stretch across.

WEB SITES

http://netcenterbu.builder.com/Graphics/Design/index.html
   Netscape's tutorial on integrating graphics with design

http://www.dsiegel.com/tips/
   Designer David Siegel's Web Wonk section on his home site; includes information about using tables for layout

http://www.quadzilla.com/tables/tableframe.htm
   Tutorial on tables: looks good and has interesting information

HOW SHOULD IT LOOK
Web Page Design

WEB PAGE DESIGN

Now that we've covered both graphics and layout grids, let's step back and look at some general principles for Web page design. Before we begin, however, let's admit something up front: Web page design has a large subjective element. Different designs will appeal to different readers (yet another reason to try to define the readers for whom you're designing your site). Yet some designs are clearly ineffective for the purpose and audience of a site, and some principles go across many design styles.

In general, you want your Web design to create a sense of unity and context for your message. Because it's so easy for your readers to slip from your site to another (particularly if you've supplied off-site links for them to follow), your readers should never be confused about whether they're still at your site. A consistent design, with some of the same elements on every page (e.g., your navigation bar, a banner at the top, a logo for your organization), will send a clear message that each page belongs to the same site.

Your design should also fit with your message: a site concerning alternative music should look very different from a site on planning for retirement. Consider carefully what colors and images go with the idea you're trying to get across.

PAGE DESIGN

Design is actually a functional part of your pages. You can use your design to create a clear hierarchy on the page, to emphasize the most important elements, and to organize your content so that it's easy to scan and understand. Design can also make your pages more pleasant to read: much better than an unbroken block of black type on a gray background.

The Grid

It's a good idea (and a time-saver) to use the same basic page layout or grid on all the pages of your site. Setting your pages up in this way has several advantages. First, it makes it easier for you to create pages quickly. Once you've got the basic grid (with, for example, a background and a table for placing your text and graphics), you can simply copy it on each page of your site and add the text and graphics that go with it. It also saves you from having to redesign each page: you simply add the graphics and text for each page you write.

More importantly, using the same page grid on all the pages in your site establishes a consistent look for your pages. Your readers will come to expect certain elements to be in certain places on your pages because they've become accustomed to your grid; you can help them both to find information quickly and to move around your site. Then they'll be less likely to be confused if they wander onto another site as to which pages belong to which author.

The grid will govern where you place your text on the page and where you place your images. If you want a left margin, for example (see the section on white space), have the same left margin on all of your pages. If you place your logo in the upper-right corner of your home page, place it there on the other pages as well. Look at the opening page of the University of Pennsylvania's history of wine site shown in Figure 6.1.

Figure 6.1 shows only the top half of the page. The bottom half has a paragraph of text explaining the site, shown in Figure 6.2.

FIGURE 6.1   Page Grid 1

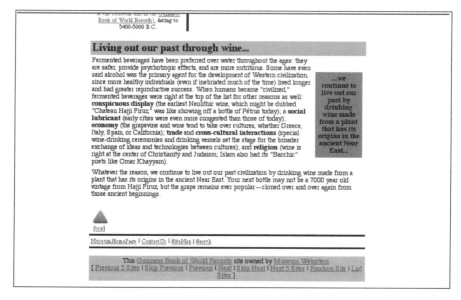

FIGURE 6.2   Page Grid 1a

Figure 6.3 shows another page from the same site on the Neolithic period. Notice that the layout of all the elements on the page is the same as Figure 6.1; the only change comes in the highlighted topic on the right and the graphic on the left.

The text at the bottom, shown in Figure 6.4, also looks exactly the same, although, of course, the content is different. In this case the text is longer, stretching to another screen, and the designer has chosen to break it up with photographs.

Finally, using a repeated page grid can also speed up your site, particularly if you're repeating graphics such as a logo or a background tile. Each time a browser opens a graphic, it loads the graphic into its memory cache; thus once the browser has loaded a graphic such as your background tile, the graphic remains in memory and won't have to be downloaded again if you repeat it on successive pages. The banner at the top of the wine page will be loaded once when you enter the site, but after that it will stay in memory.

■ TRY IT

Go to a Web site you like and look at several pages. Does the site use a grid? What elements repeat from page to page? How does the site create unity in its design? Bring your examples to class.

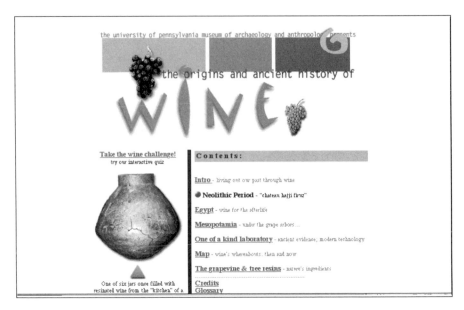

FIGURE 6.3 Page Grid 2

Top of the Page

In many ways the top of your page is the most important part, similar to the above the fold part of a newspaper page. It's the part of the page that everyone will see first, and in many cases it will make people decide to look further or

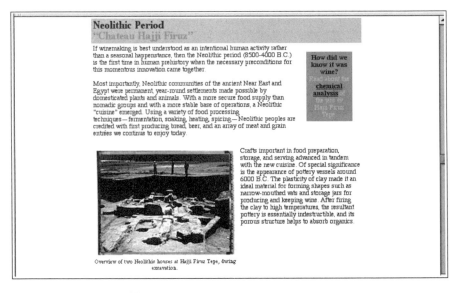

FIGURE 6.4 Page Grid 2a

move on. You do want something at the top of the page that identifies your site and sets the prevailing mood and style. However, beware of the large, long-loading graphic! Your readers may not stick around if it takes too long for the top of the page to come up.

Look again at the banner at the top of Figure 6.1; although it's relatively simple, it effectively gives the identity of the site and sets the graphic style of the site as a whole. In his discussion of Web design for CNet, Ben Benjamin argues that everything the reader needs should be found within the first 300 pixels of the page; if you've got a nonscrolling reader, that's certainly going to be the case!

Page Dimensions

It seems self-evident that the computer screen is smaller than the page, and the browser window is smaller still. But sometimes it's hard to remember that you have limited space in which to work. The default browser window size is usually given as 480 by 640 pixels (on higher-resolution computers, 600 by 800); however, that doesn't mean you should design your graphics and your pages to fill those dimensions. The *Yale Style Guide* gives a "graphic safe area" of 295 pixels high by 595 pixels wide; this is the area that will come up in most browsers without scrolling. Macromedia, Inc. enlarges this somewhat in its Dreamweaver program to 300 by 600 pixels, but that's still pretty close.

It's particularly important to keep your graphics to these dimensions; otherwise you'll force your readers to scroll both horizontally and vertically to see your entire page—not a good idea!

DESIGN PRINCIPLES FOR THE WEB

So far we've discussed basic elements of an effective page—a repeating page grid that places important information and identifiers at the top of a 600-pixel by 300-pixel screen. But how do you go about designing those elements? In his book *Click Here,* designer Raymond Pirouz identifies several basic elements that Web design has in common with other page designs, including color, white space, contrast, scale, and typography. Each of these elements, singly and in combination, can be used to produce an effective design; used badly or ignored, they can undercut your site's effectiveness.

Color

As explained in Chapter 2, the palette of colors available for use on your Web site is more limited than the palette you can use in print—255 colors rather than millions of shades. However, using the same set of colors throughout your site can help to create the sense of unity and context that you want to continue from one of your pages to the next. It's a good idea to use the same basic color scheme on all of your pages; shifting from one set of colors to another may make readers think they've inadvertently wandered into another site.

Choose your colors from the browser-safe palette you downloaded in Chapter 2, and consider the colors that are most appropriate for the topic of your site. Bright colors work well for energetic, jazzy sites. Cooler, more conservative shades will work better with more conservative sites. If you have lots of text, make sure your background colors are pale: white, cream, or pastel. The Palette Man site (http://www.paletteman.com/) will let you compose a palette of browser-safe colors and test them together to see how compatible they are. Project Cool (http://www.projectcool.com/developer/reference/color-chart.html) will let you look at a color used both as text and as background.

You can also use color to draw attention to elements you want to emphasize. A red logo will draw attention immediately; colored headings will stand out. You can even use color to emphasize text that you especially want readers to see. Just be careful not to overdo it: too much emphasis has the same effect as no emphasis at all—if everything is emphasized, nothing stands out.

White Space

Actually, of course, white space doesn't have to be white: it's unfilled space on the page. White space can be used to direct the eye and to help the reader understand how information is organized. For example, a heading on a line by itself will automatically draw the eye more than a heading on the same line with text. The white space surrounding the heading gives it added emphasis; we immediately assume that this kind of heading is more important than the heading without white space because the space helps to reinforce its importance.

One way to use white space effectively on your Web pages is to limit the width of your text. Text that flows from one side of the page to another, as in Figure 6.5, is difficult to read.

We're used to text with margins; in fact, many hard-copy texts now use heads in a separate column at the side (side heads), which introduces even more white space. You can set up margins on your page by using tables for layout and creating empty cells for white space, although you can also set margins using Cascading Style Sheets (see Chapter 9). To add a margin to Figure 6.5,

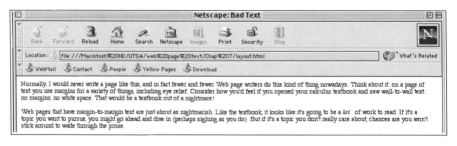

FIGURE 6.5   Wide Text Lines

use a two-column table with a nonbreaking space in the column on the left and a narrower text column on the right. The HTML would look like this:

```
<HTML>
<HEAD>
        <TITLE>Bad Text</TITLE>
</HEAD>
<BODY BGCOLOR="#FFFFFF">
<TABLE BORDER="0" CELLPADDING="0"
CELLSPACING="0">
<TR><TD WIDTH="100"> </TD>
<TD WIDTH="500">
Normally, I would never write a page like this,
and in fact fewer and fewer Web page writers do
this kind of thing nowadays. Think about it: on a
page of text you use margins for a variety of
things, including eye relief. Consider how you'd
feel if you opened your calculus textbook and saw
wall-to-wall text: no margins, no white space.
That would be a textbook out of a nightmare!

<P>Web pages that have margin-to-margin text are
just about as nightmarish. Like the textbook, it
looks like it's going to be a <I>lot</I> of work
to read. If it's a topic you want to pursue, you
might go ahead and dive in (perhaps sighing as
you do). But if it's a topic you don't really
care about, chances are you won't stick around to
wade through the prose.</TD></TR></TABLE>
</BODY>
</HTML>
```

In a browser, it looks like Figure 6.6.

To make that text even easier to use, you could place side headings in the empty column on the left, which would look like Figure 6.7.

The changed line of code for the table is this:

```
<TD WIDTH="100" VALIGN="Top"><H3>Using the Right
White Space</H3></TD>
```

Notice how the heading is emphasized by the white space that surrounds it. Notice how it also leads your eye to the beginning of the text block. That's what white space can do: it emphasizes and gives shape to your information. As Pirouz puts it: "White space can be described metaphorically as a river, leading the user through fields of information" (p. 39).

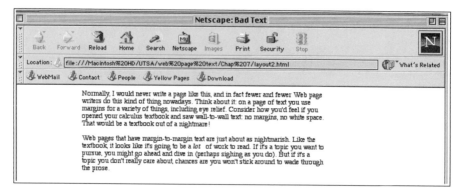

FIGURE 6.6   Text on Table

■ TRY IT

Write the HTML and text for a page that has a left margin of 100 pixels. Remember to include the basic template: `<HTML></HTML>`, `<HEAD></HEAD>`, `<TITLE></TITLE>`, `<BODY></BODY>`

If you're working on your text-based site, you can set up margins to help organize your information.

Contrast

Your eye is naturally drawn to anything on a page that's different from everything else: that's the power of contrast. In Figure 6.7, the heading contrasts

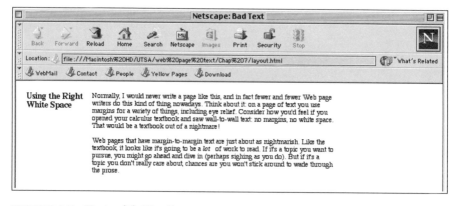

FIGURE 6.7   Text with Heading

with the text: it's darker, larger, and surrounded by white space. It's a good idea to figure out what elements on your page should stand out: what elements do you want to emphasize by increasing their contrast? In his book *Designing Visual Interfaces,* Kevin Mullet, of Macromedia, identifies seven qualities that can be used to create contrast: "shape, size, color, texture, position, orientation, and movement" (p. 52). How do they work? Here are some examples:

- *Shape:* An irregularly shaped graphic will contrast with a rectangular text block.
- *Size:* A larger heading will contrast with a smaller line of text.
- *Color:* A blue link will stand out against a black sentence.
- *Texture:* A button with a patterned texture will contrast with a plain background.
- *Position:* A logo placed high on the page will draw attention more quickly than a button placed midway down.
- *Orientation:* A graphic that's placed vertically will contrast sharply with a square of text.
- *Movement:* An animation (either an animated GIF or a more complex image) will automatically draw your attention over something static.

Let's see how this works in practice.

There are several uses of contrast on the home page for the Messy Gourmet site in Figure 6.8. The designer uses *shape* to contrast both the flower on a stain logo and the word *Messy,* both of which have irregular outlines compared

FIGURE 6.8 Page Using Contrast

to the rectangular text blocks. *Size* contrasts are used both with the Messy Gourmet banner at the top and the section headings below (i.e., Manifesto, Cool Gadgets, etc.), contrasted with the smaller text font used beneath them. *Color* may be difficult to see in a black-and-white reproduction (to see the full-color version, visit the site at http://www.messygourmet.com/), but the text is a cool violet on a white background, whereas the links are green and the icons are shades of yellow, as well as highlight and accents in green and violet. *Texture* contrasts are created by the irregular splotches around *Messy* compared to the smooth regularity of *Gourmet*. The top *position* of the banner gives it extra emphasis by surrounding it with white space (and the top of the page is a naturally emphatic position). This emphasis is heightened by the contrast in the *orientation* of the word *Messy:* whereas everything else is very symmetrical in the layout, *Messy* seems to sprawl asymmetrically across parts of both columns.

■ TRY IT

Visit a Web site whose design you like. What are the elements of contrast and how are they used? How does the designer make contrast a part of the page layout? Bring your answers to class.

Scale

Scale is another means of drawing the reader's attention. If everything on a page is the same size, everything will get equal attention—nothing will stand out. But if size is used according to importance—that is, if the elements on your page are sized to indicate their place in the hierarchy of information—your reader can deduce the relationship of those elements at a glance. Scale works particularly well if you couple it with another means of emphasis, such as color. A large, brightly colored element will attract attention even if it's placed lower on the page.

You can see the effects of scale on the Messy Gourmet page. Notice how the size of the headers automatically indicates that they represent major sections: they draw your eye immediately, as soon as you look down from the banner. Now consider how that would change if the icons beside the headers were larger. As it is, the icons are small enough not to distract your eye from the headings, where your main attention should be placed. If they were bigger, their size might throw off the scale of the page.

■ TYPE TERMS ■

Display fonts: Type fonts designed to be used for stylistic effects in large sizes (e.g., Anna, Bergell, Homely, Umber, etc.).

Font: A complete set of letters in a given type style (e.g., Times, Helvetica, Futura, etc.).

Leading: The space between lines of type, measured in points.

Sans serif: "Without serif." Fonts that have no serifs at the ends of letters.

For example, Helvetica: **S**

Serif: Short extensions at the ends of letters in some fonts.

For example, Times: **S**

Screen font: A font that has been designed especially for use on computer monitors (e.g., Arial, Chicago, Geneva, Georgia, Verdana).

Typography

The use of type on the Web is changing due to Cascading Style Sheets (CSS) (see Chapter 9), and typography will change even more radically in the near future when programs that embed type fonts on Web pages become more available. However, at the moment the use of type on the Web faces a huge roadblock: the text type that appears on a Web page is governed by the fonts that are loaded on the reader's computer. This means that even if you use CSS to specify that your page should be shown in twelve-point Palatino, you'll get your wish only if your readers have twelve-point Palatino loaded on their machines. Otherwise, they'll see your pages in another serif (see the Type Terms box) font, probably Times.

One way around this problem is to create type in a graphics program such as Adobe Photoshop. Although you wouldn't do this for all the type on your page (think how large the graphic would be!), you can create type for headings and for banner effects. (That's how the headers on the Messy Gourmet page were created.) After you've created your type in your graphics program, save the type as a GIF and place it on the page as you would any other graphic.

Type can be a powerful means of establishing a visual style and mood, because type fonts (particularly display type) have distinct characters. Consider these examples:

ANNA Bergell STOP

Goudy Harrington

Lucida Skia

Tekton Umber

Each of these fonts creates a definite style and mood; you might use Anna for sophistication, Bergell for casual, and Umber for a Celtic feeling. Using words themselves as graphics can give your site a unique flavor, and at a resolution of 72 dpi, most word–graphics will be relatively small. Consider how fonts are used for stylistic effects on the site in Figure 6.9.

The site name, *The Diego Rivera Mural Project,* as well as *City College of San Francisco* and the language choices, are all done using an art deco font that fits historically with the period of the Rivera murals. In addition the curves and slants of the letters (look particularly at the lowercase *e*) seem to pick up the curves of the mural panel; the designer has also used a series of thin lines to tie the text block together and reinforce the art deco elegance of the site. The Messy Gourmet is another site that uses fonts effectively: look at the contrast in fonts between the words *Messy* and *Gourmet* in Figure 6.8, for example. The *Messy* font looks just that!

■ WHICH FONT? ■

You're much more limited in choosing a font for your Web pages than in choosing a font for a print project; generally, you have your choice of Times, Courier, and Arial or Helvetica (Arial is a Windows True Type font, but it's also found on many Macs). Cascading Style Sheets will allow you to expand your font choices somewhat, but they'll still be limited to what your readers have loaded on their computers. So which of these fonts is best? Many designers have argued that the serifs (see the Type Terms box) on serif fonts (which may help to make serif fonts more readable in print because they can help to guide the eye along the type line) are frequently lost in the lower resolution of computer monitors. Thus Times, the default Web choice, may not be the best choice if you have extended type blocks. In this case you may be better off spec-

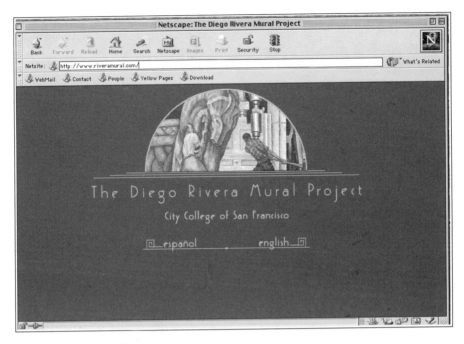

FIGURE 6.9   Use of Fonts

ifying a sans serif font such as Arial or Helvetica. Many screen fonts are sans serif (e.g., Geneva for Mac, Arial for Windows, Microsoft's Verdana). You can specify your font by either using the Cascading Style Sheets method of designating fonts (see Chapter 9) or, if all else fails, using `<FONT FACE="Arial">` at the beginning of your tags. (`FONT FACE` doesn't work in all browsers and browser versions, although it's used by many authoring programs.)

Type can also be used to create a consistent and appealing page design that will help your readers locate information quickly. If you set up regular patterns of type use on your page—that is, headings, subheadings, and text, separated by regular use of white space—you help your readers scan your pages before they begin reading. You should decide how you want your text and your headers to look and then maintain the same design from page to page. You can create a style sheet for yourself, indicating what font size, color, and treatment (e.g., italic, boldface) you'll use for each element on your page, including text, headings, and links. You can later formalize this style sheet using Cascading Style Sheets if you wish.

Another thing to be aware of with type is the length of your text line. As mentioned in the earlier discussion of white space, lines of text that spread all the way across the browser window may be too wide to be read easily; in fact, some designers argue that lines should contain no more than twelve to fifteen words. Magazine and book columns are kept narrow to accommodate the eye's span of movement—eight centimeters (around three inches). According to the *Yale Style Guide,* the average Web page is around twice as wide as the viewer's eye span. Thus you'll need to take extra care to make sure your lines of text aren't too wide for your readers to read comfortably. Avoid dense, forbidding blocks of words by limiting the width of your line, using tables or **BLOCKQUOTE**.

▪ TRY IT

Visit a Web site that you like. What kind of type does it use? Does the designer use type for graphic effects as well as text? Is the use of type effective? Bring your answers to class.

▪ TABLES AND TYPE ▪

Tables present some special difficulties for typography. Remember, the **FONT** or **BASEFONT** tags you use for the text on your page won't affect the type you place on your layout tables. If you want to change the type size or color or font on your tables, you'll need to enter the changes after the **<TD>** or **<TH>** tags. So, for example, if you wanted to use a slightly larger text font in a contrasting color on your table, the setting would be

```
<TD><FONT SIZE="4" COLOR="red">Text would go here.
</FONT></TD>
```

MULTIMEDIA EFFECTS

One of the many features that make the Web different from print is the fact that Web pages can use multimedia effects. Your pictures don't have to stand still: you can import various kinds of movement as well as sound.

Movement is the ultimate attention getter: a moving image will automatically overwhelm every other element on your page. However, any moving image will take up more memory than a static one: make sure that the image is worth the burden. It's infuriating to wait for several minutes while an animated GIF loads, only to discover that it's something cheesy and irrelevant to the page. (The days when all animated images had a gee whiz effect are long past!) In fact, relevance to the page should be the central issue. If an animation will help to accomplish your purpose (e.g., rotating an object so that a reader can see both sides), then feel free to include it. But if the animation is there only to supply glitz, it's best to forget it.

You can create your own animations using GIF images and shareware programs, such as GIF Animator; some commercial imaging programs, such as Adobe's Image Ready and Macromedia's Fireworks, also include this capability. You can also download GIF animations from the Web. The tag for inserting GIF animations is the same as the one for inserting static images: that is, `IMG SRC="animated.gif"`. Figure 6.10 shows an example of a site that uses movement effectively.

The Smarty Pants Yo-Yo site presents a new animation of a yo-yo trick every week. Here the animation is worth waiting for because it reinforces the content of the site.

Sound can also be used to add effects to your site, but you must be careful about copyright issues: you can't just add your favorite cut from the latest CD. In fact, copyright law limits the number of seconds of a song that can be used without a copyright violation. You should also consider the cumulative effect of sounds. If the cursor beeps every time it passes over a certain point on the page, your readers may start avoiding that part of the page after they get tired of hearing it! In general, treat sounds as you treat movement—if the sound helps you to accomplish your purpose, you can include it. But skip the extraneous effects!

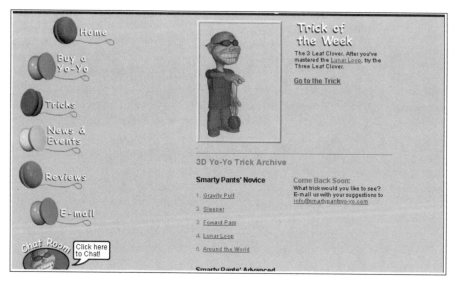

FIGURE 6.10 Multimedia Effects

A variety of sound technologies are available now, from simple recordings using the utilities present with Windows and Mac OS to sophisticated streaming audio that plays as it downloads large files. Each format will use a different extension and some use different loading and coding formats. If you want to do more with sound, consult one of the many books or Web site tutorials available on Web sound design.

■ TRY IT

Go back to the clip-art sites you visited in Chapter 3. Which ones also offer animations and/or sound files? Are they effective or are they extraneous? Bring your answers to class.

FRAMES

Frames offer a way to divide Web pages into multiple, independently controlled sections, each using an independent HTML file. Using frames can make it possible for you to have certain features in the browser window that stay visible on every page, such as a menu of pages available on your site or a banner or an index of information, while other parts of the window change when you click a link. For example, the Smarty Pants Yo-Yo site in Figure 6.10 uses frames to place the menu on the left of the page on all the pages within the site while other parts of the browser window change, depending on the choices made on the menu.

Frames offer many advantages for layout: they're very good for making comparisons between items because you can have items side-by-side in the frames. They can break your page into interesting segments that you can treat individually: you can put a banner in a frame, for example, without having it influence the material in the other frames, as it would if you placed it on the page without a grid. Frames also allow your reader to have some interaction with your page; as the reader clicks a link, a frame on the page shows the changes.

However, frames also have a downside. For one thing, they can fragment your page, making it hard to see how the various parts of the window are related. Users sometimes become lost in pages with frames, particularly if it isn't clear how the frames work together, or if the frames all seem to operate independently. Finally, because frames require more than one HTML file to be read, they can take more time to load than a page without frames.

If you're going to use frames, you should sketch out the frame layout in advance, trying not to get so many frames within a window that the user (or the writer) becomes confused. Make sure the frames are clearly interrelated, and that the reader can see what the relationships are.

Frames are somewhat complex and time-consuming to code; this is one situation in which using an authoring program may have real advantages. On the other hand, by now you should have a sense of why it's useful to know what the HTML conventions are for a Web page element, so instructions for basic frame layout are provided in the Appendix at the end of the book if you'd like to try doing them yourself.

PUTTING IT ALL TOGETHER

You can try out your page designs in your graphics program, creating a mock-up of your page grid to see how it will look on a computer screen. Most graphic programs will also allow you to create graphics at precise sizes, so you can make sure you're designing at no more than 300 by 600 pixels.

When it comes to the actual design of your pages, you'll be guided by your purpose and audience, as you are when you create any page. However, remember the basic principles:

- Your readers should be able to see the relationships between your topics from the way that they're placed on the page. Important points should be emphasized visually, and the progression down your page should have a logical order.
- Simple is usually best in Web design: too much glitz can overwhelm your content. If you use multimedia, make sure that it's relevant to your topic and worth the added memory demands it will place on your site.
- Keep your design consistent from page to page, so that your readers can learn to use your site as they move around it.
- Using "redundant" page design not only will help your readers to use your site but also will make your pages more efficient and faster loading.

The best advice for coming up with a design for your site is to spend some time on the Web. Look at sites that deal with similar topics. How are they designed? Do they appeal to you? Are they effective? If not, what's the problem and how could you fix it? Keep track of sites that you particularly like, even those that are on different topics. Have the designers of these sites employed effects that you'd like to use? How can you adapt some of the elements you like to your own designs?

One of the best ways to develop effective Web sites is to look at other effective Web sites and adopt some of their practices. Look at the HTML for pages you like (in Netscape, go to the View menu and select Page Source; in Internet Explorer, go to the View menu and select Source). If you're curious about how a designer achieved a particular layout, check out the code and see how it was created.

The best way to design effective pages is to see what works, and then to try to re-create it. It may take a few tries, but you'll eventually come up with something that's right for you.

■ WRITING ASSIGNMENT

Design a Web page using a page grid, white space, and Pirouz's design principles. Your page should include at least *three* paragraphs and at least *one* graphic.

Tips.   If you're working on the text-based site assignment, try creating a grid that you can use on the entire site. Use this page as an opportunity to set up a design that's both attractive and functional—one that will unify your site. Or you can start working on the image-based site assignment given in Chapter 8.

WEB SITES

Example Sites

http://www.messygourmet.com/
   The Messy Gourmet cooking site

http://www.upenn.edu/museum/Wine/wineintro.html
   University of Pennsylvania wine history site

http://www.riveramural.com/
   Rivera murals site from City College of San Francisco

http://www.smartypantsyo-yo.com/main/index.html
   The Smarty Pants Yo-Yo trick of the week

Reference Sites

General Design

http://info.med.yale.edu/caim/manual/pages/page_design.html
   Yale Style Guide section on page design, including grids giving dimensions for browser windows

http://wdvl.internet.com/Authoring/Design/Pages
   Tutorial on Web design by Charlie Morris at the Web Developer's Virtual Library

http://www.builder.com/Graphics/Design/index.html
   CNet's tutorial on Web design by Ben Benjamin

Web Typography

http://info.med.yale.edu/caim/manual/pages/typography.html
  Yale Style Guide section on typography

http://wdvl.internet.com/Authoring/Design/Pages/typography.html
  Charlie Morris on typography

http://webreview.com/wr/pub/Fonts
  New from Daniel Will-Harris: a magazine on Web typography

http://www.builder.com/Graphics/Design/ss2d.html
  CNet's Ben Benjamin on typography

http://www.will-harris.com/type.htm
  Typofile. Idiosyncratic type magazine with info. about Web screen fonts

Page Length

http://info.med.yale.edu/caim/manual/pages/page_length.html
  Yale Style Guide discussion of page length

http://www.sun.com/styleguide/tables/Page_Length.html
  Sun Microsystems Guide to Web Style discussion of page length

Color

http://www.paletteman.com/
  The Palette Man site, which lets you try out different combinations of Web-safe colors

http://www.projectcool.com/developer/reference/color-chart.html
  Project Cool's color chart that lets you see how a color looks used as text and as background

Design Tips

http://bignosebird.com/fdoor.shtml
  Big Nose Bird tutorial on creating an effective front door (i.e., first impressions count)

http://home.netscape.com/computing/webbuilding/studio/feature19980827–3.html
  "Designing Words for the Web" by writer–designer Derek M. Powazek, including information about making text more readable

http://www.design.ru/ttt/index.html
  Russian designer Dmitri Kirsanov's "Top Ten Tips" for Web page style, idiosyncratic and great fun, although a little dated by now

http://www.htmlgoodies.earthweb.com/
Designer–educator Joe Burns's site includes information about preparing and mounting sound files

http://www.wpdfd.com/wpdhome.htm
Web Page Design for Designers: a magazine focusing on Web design

CHAPTER 7

MAKING THE WEB A WEB
Links

WHAT MAKES THE WEB A WEB?

Webs by their very nature are complicated structures of interconnections; even a spider's web is so intimately connected that a vibration in one area will be sensed by the spider sitting in another area. The World Wide Web is no less interconnected, and links are what make those interconnections work.

Technically, a *link* is a connection between two Web pages, sometimes pages on the same Web site and sometimes pages on different sites. You can also have *anchor* links on a single page, such as the links that allow you to click back to the top of a long page without scrolling. However, at the base of most links is either an absolute or relative *URL* (uniform resource locator).

URLs

URLs are addresses for Web pages; they indicate the location of a given page or a given site. In other words, the URL tells the browser where to look to find a particular Web page. Let's look at a typical URL:

```
http://www.cofah.utsa.edu/mbatch/default.html
```

http: http tells the browser what protocol to use when it requests the page from a server (`http` stands for hypertext transfer protocol). Most Web pages use `http` rather than other protocols such as ftp or gopher. Not all browsers require `http` in your address, but it doesn't hurt to include it.

www This section tells the browser that the page is located on the World Wide Web, the section of the Internet that uses HTML. Again, not all browsers require `www`, but including it will help to ensure that the page will be found.

cofah.utsa.edu This section of the URL is called the *domain*; it tells the browser what server to contact for the page. In this case the name of the server is `cofah` (College of Fine Arts and Humanities), and it's located at `utsa` (the University of Texas at San Antonio), an educational institution (hence, `.edu`).

/mbatch/default.html   The rest of the address gives the path that the browser must take to find the particular page requested. `mbatch` is a directory found on the `cofah` server. The order of directories in the URL goes from higher (more inclusive) to lower—`cofah` includes the `mbatch` directory, which includes the file `default.html`. The final part of the URL, `default.html`, is the actual HTML file the browser is to read.

This particular URL is *absolute;* that is, it tells the browser every part of the address so that the browser can locate that page exactly. However, sometimes you don't have to use the entire URL—just a part of it. Particularly when you're linking together the pages you've created on your own site, you usually don't need to supply the protocol and domain names because the browser is simply moving from one page to another in the same domain using the same protocol. These shortened URLs are called *relative* URLs, because they're relative to other pages on the same site. So, for example, on the page listed above (at the university UTSA), a link to another page, `welcome.html`, is part of the same Web site as `default.html`. The link to this second URL is written as `welcome.html` rather than `http://www.cofah.utsa.edu/mbatch/welcome.html` because its address is relative to the address of the page the browser has already opened. By linking to `welcome.html`, you're telling the browser to look for this page in the same directory (i.e., on the same Web site) where it found `default.html`. Relative URLs are discussed further later in this chapter when we talk about linking your pages.

Relative URLs are used to move from one page of a site to another and to locate resources such as graphics. (A graphic name such as `idea.gif` is actually a relative URL.) They may include names of directories along with the name of the file. For example, if all of your graphics are together in a directory called `images`, then the relative URL for one of these graphics would be `/images/idea.gif`. However, relative URLs won't include the complete address; that is, they won't repeat the complete protocol, domain, and path designations.

You'll use a URL when you write most of your links; the only exception will be when you write anchor links on the same page.

LISTS

Before we get to the actual HTML for linking, let's look briefly at the way HTML handles lists because you'll frequently use lists to present your links. HTML has three types of lists: ordered, unordered, and definition.

Ordered Lists

Ordered lists use numbers. You can use an ordered list when you want to maintain the items you're listing in a particular order, for example, a series of steps or a list of links where some of the links are more important than others. Ordered lists begin and end with container tags: <OL></OL>. Then within the containers, you use a List Item tag <LI> to mark each item on the list. When you look at the list in the browser, the items will be numbered in order. This is the HTML:

```
<HTML>
<HEAD>
       <TITLE>Lists</TITLE>
</HEAD>
<BODY BGCOLOR="#FF3366" TEXT="FFFF00"> <FONT
SIZE="4">
<OL>
       <LI>This is the first item on my list
       <LI>This is item two
       <LI>This is item three
       <LI>I could go on, but you get the idea
</OL></FONT>

</BODY>
</HTML>
```

In a browser it looks like Figure 7.1.

Creating an Ordered List

1. Type any introductory text for the list (e.g., a title).
2. Type <OL>
3. Type <LI> and the text for the first list item.
4. Repeat Step 3 for each of the other items on the list.
5. Type </OL> to complete the ordered list.

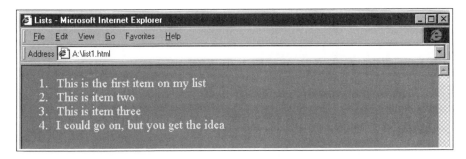

FIGURE 7.1   Ordered List

■ TRY IT

Write the HTML and text for a page containing an ordered list. The list should include at least three items. Remember to include the basic template: <HTML></HTML>, <HEAD></HEAD>, <TITLE></TITLE>, <BODY></BODY>

If a list fits your purpose, you can incorporate this page into your text-based or image-based site.

Unordered Lists

Unordered lists can be used whenever the order of the items on the list doesn't matter, such as a shopping list or a list of reminders. In its default form, the unordered list uses bullets, and the HTML is very similar to that used for ordered lists. The container tags are <UL></UL>, and within the container, you use the same list item tag, <LI>. The HTML looks like this:

```
<HTML>
<HEAD>
        <TITLE>Lists</TITLE>
</HEAD>
<BODY BGCOLOR="#FF3366" TEXT="FFFF00"> <FONT
SIZE="4">
<UL>
        <LI>This is the first item on my list
        <LI>This is item two
        <LI>This is item three
        <LI>I could go on, but you get the idea
</UL></FONT>

</BODY>
</HTML>
```

Figure 7.2 shows what it looks like in the browser.

■ CREATING UNORDERED LISTS

1. Type any introductory text for the list.
2. Type <UL>
3. Type <LI> and the text for the first list item.
4. Repeat Step 3 for each of the other items on the list.
5. Type </UL> to complete the unordered list.

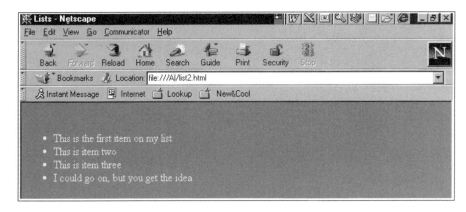

FIGURE 7.2   Unordered List

■ TRY IT

Write the HTML and text for a page containing an unordered list. The list should include at least three items. Remember to include the basic template: `<HTML></HTML>`, `<HEAD></HEAD>`, `<TITLE></TITLE>`, `<BODY></BODY>`

If a list fits your purpose, you can incorporate this page into your text-based or image-based site.

You can use other symbols such as circles or squares in the unordered list besides bullets. To change the symbol, add `TYPE` to `UL` for the entire list or `LI` for individual list items; `TYPE="circle"` will produce a circle and `TYPE="square"` will produce a square. (`TYPE="disc"` will produce a bullet, but you don't need to include `TYPE` for that because it's the default style.)

The HTML looks like this:

```
<HTML>
<HEAD>
        <TITLE>Lists</TITLE>
</HEAD>
<BODY BGCOLOR="#FF3366" TEXT="FFFF00"> <FONT
SIZE="4">
<UL>
        <LI>This is the first item on my list,
using a bullet
        <LI TYPE="circle">This is item two (a
circle)
```

```
        <LI TYPE="square">This is item three (a
square)
</UL></FONT>

</BODY>
</HTML>
```

In a browser, it looks like Figure 7.3.

Definition Lists

The last type of HTML list uses a term and definition layout; there are two parts—a term, which appears at the left margin, and a definition, which is indented. Actually, you can use the definition list format when you want an indented list but not the bullets. The list container is <DL></DL>. The tag for the Definition Term is <DT>; the definition (or definitions) tag is <DD>. The code looks like this:

```
<HTML>
<HEAD>
        <TITLE>Lists</TITLE>
</HEAD>
<BODY BGCOLOR="#FF3366" TEXT="FFFF00"> <FONT
SIZE="4">
<DL>
        <DT>This is a definition list
                <DD>And these are the indented items
                <DD>I don't have to stop with just one
                <DD>If I want an unbulleted list, this
format works well.
</DL>
</FONT>

</BODY>
</HTML>
```

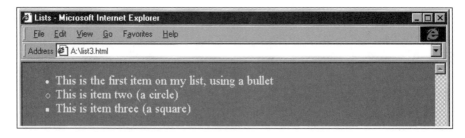

FIGURE 7.3   Unordered List with Symbols

Figure 7.4 shows what it looks like in the browser.

Creating Definition Lists

1. Type the introductory text for the list.
2. Type <DL> (Any text you enter after this marker will appear on its own line, aligned to the left margin.)
3. Type <DT>
4. Type the term or phrase to be defined or explained.
5. Type <DD>
6. Type the definition or explanation of the term in Step 4.
7. Repeat Steps 3 through 6 for all the terms on the list.
8. Type </DL>

■ TRY IT

Write the HTML and text for a page containing a definition list. The list should include at least three items. Remember to include the basic template: <HTML></HTML>, <HEAD></HEAD>, <TITLE></TITLE>, <BODY></BODY>

If a list fits your purpose, you can incorporate this page into your text-based or image-based site.

■ LIST TIPS ■

- Keep the text in your list items short.
- Inserting a line break <BR> in a list item will break the text to the next line but maintain the indenting.
- Text placed after the <OL> or <UL> marker without the <LI> will appear indented like a list item but without the number or symbol.
- To change the symbols used in ordered and unordered lists, use the following (the default is 1, 2, 3, etc., and ●):

FIGURE 7.4 Definition List

- After <OL, type **TYPE=** and one of the following symbols: **A** for capital letters, **a** for small letters, **I** for capital roman numerals, **i** for small roman numerals, or **1** for numbers—the default.
- After <UL, type **TYPE=** and one of the following words: **disc** for a solid round bullet; **circle** for an empty round bullet; or **square** for a square bullet.
- You can change the first number in an ordered list (if you want to begin somewhere other than 1); after <OL, type **START=** and the number you want to begin with.

MAKING LINKS

The HTML for both absolute and relative links is basically the same, although the URLs are written differently. There's one thing to remember, though: as with your graphic tags, when you create a link, you must type the URL *exactly* as it's written on the page you're linking to, including capitalization and spelling.

Absolute Links

All links use the same container tags **<A HREF="url"> </A>**; the difference comes in what you type into that **"url"** space. With absolute links (which include most links off your site to other Web sites), you must use the entire URL, including protocol, server, and path. For example, let's say you wanted to link to the Webopaedia site (an online encyclopedia of Web terminology). The first part of your link would read: **<A HREF="http://www.pcwebopaedia. com">**. You don't need to include an html file name here because the **www. pcwebopaedia.com** designation links to the entry page of the site. Notice that the URL is **pcwebopaedia** rather than webopaedia. Although the latter is the name used on the site, **pcwebopaedia** is the *domain* name. You must be careful to write the domain name exactly as it's given in the site URL. If you've ever spent some time trying to guess what the correct URL for a site is without knowing the exact domain name (maybe typing in dozens of variations on the name of a company), you know how important it can be to get the domain name exactly as it appears on the page to which you're linking.

Clickable Text

Between the **<A HREF="url">** and the **</A>**, you'll write whatever you want your users to click on to activate your link. This can be either text or a graphic. (We'll discuss using graphics for links a little later in the chapter.)

Whatever text you place between the two container tags will probably be underlined in the browser because underlining is the way in which most browsers indicate links. (This is a good reason not to use underlining for anything else on your site!)

Some things to keep in mind when you're creating your clickable text:

- Keep the text to be clicked relatively short. (It's hard to click on large segments of text.)
- Don't use "click here" for your text; users need to have some idea of where they're going to go if they click on the link.
- If you use italics or boldface on the clickable text, be consistent from one link to another. (That is, don't italicize some links but not others: you'll only confuse your readers.)

Creating Absolute Links

1. Put the cursor on your HTML page where you want the link to appear.
2. Type `<A HREF="url address">`, filling in the complete address of the page you want to link to.
3. Type the clickable text that will be underlined or otherwise highlighted so that the user can click on it and initiate the link.
4. Type `</A>`

▪ TRY IT

Create the HTML and text for a page that links to a page on another Web site. Be sure to include the complete URL for your link. Remember to include the basic template: `<HTML></HTML>`, `<HEAD></HEAD>`, `<TITLE></TITLE>`, `<BODY></BODY>`

You can add absolute links to either your text-based or image-based site—wherever these links are appropriate.

Link Colors

You can change the colors of your links if you wish. (Many designers argue that if you change one color on your site—e.g., background or text—you need to change every color to avoid confusing users who have reset their default colors.) All of the color changes are attributes within the `BODY` tag. To change the color of your links, use `LINK="either a hexadecimal number or a` color name". To change the color of links that the user has already visited, use VLINK="either a hexadecimal number or a color name".

As for what color to make your links, in general you want your links to stand out; thus you should make them a color that contrasts with the color of your text. Some designers also argue that if you make your links a "hot" color, such as red or orange, it will imply that the links are lively. Using this reasoning, you'd also make the visited links a cool color, such as blue or green. The downside of this idea, however, is that the Netscape default link colors are blue for links and red for visited links; thus experienced Web surfers may be accustomed to seeing links as cool colors and visited links as hot colors. In general, choose your link colors to complement the colors of your site, and if you change colors for text and background, it's best to change link colors as well.

If I put all of these options together, the HTML looks like this:

```
<HTML>
<HEAD>
      <TITLE>Links</TITLE>
</HEAD>
<BODY TEXT="#0033FF" VLINK="#33CC33"
BGCOLOR="#FFFF00" LINK="#FF6633"
ALINK="#FF0033"><BASEFONT SIZE="4">
<OL>
<LI><A HREF="http://www.facstaff.bucknell.edu/
rbeard/diction.html"> Web of Online
Dictionaries.</A> A multilingual dictionary site
that allows you to look up words in several
different languages through links to other
online dictionaries.
<LI><A HREF="http://www.notam.uio.no/~hcholm/
altlang/"> Dictionary of Alternative
Languages.</A> A multilingual slang dictionary
compiled collaboratively by Internet users.
<LI><A HREF="http://www.spellweb.com/">
SpellWeb.</A> If what you want is a correct
spelling, this site will compare two alternative
spellings and tell you which one is more popular
(according to Internet users).
<LI><A HREF="http://www.pcwebopaedia.com/
">Webopaedia.</A> Dictionary of computer
terminology with extensive links for each term.
</OL>

</BODY>
</HTML>
```

In the browser, it looks like Figure 7.5.

Changing Link Colors

- To change the link color, within the `BODY` tag, type `LINK="#RRGGBB"` (a hexadecimal color number or one of the predefined color names).
- To change the color of visited links, within the `BODY` tag, type `VLINK="#RRGGBB"` (a hexadecimal color number or one of the predefined color names).

■ TRY IT

Using the page with the link that you worked with in the previous exercise, change the link colors.

Linking to E-mail

Most Web pages include an e-mail address for someone associated with the page: the president of the organization, the manager of the company, the sales office for the product. If nothing else, most pages include the e-mail address of the designer, frequently in a separate address section at the bottom of the first page.

Linking to your e-mail address uses a format very similar to that of absolute links, with one major difference: instead of using an absolute URL, you insert `"mailto:designer@emailaddress.com"`, with your e-mail address placed after the colon.

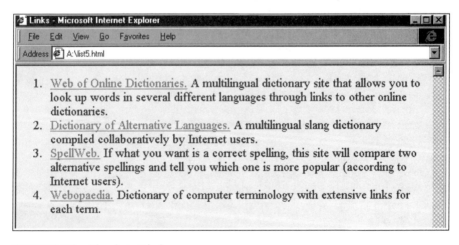

FIGURE 7.5 Absolute Links

Linking to an E-Mail Address

1. At the point where you want your address to appear, type `<A HREF= "mailto:e-mail address">`, giving your e-mail address after the colon.
2. Type the clickable text to indicate that this is your e-mail address.
3. Type `</A>` to complete the definition of the link.

■ TRY IT

Add your e-mail address to the page worked with in the previous exercises.

Adding an e-mail address to the previous HTML looks like this:

```
<HTML>
<HEAD>
        <TITLE>Links</TITLE>
</HEAD>
<BODY TEXT="#0033FF" VLINK="#33CC33"
BGCOLOR="#FFFF00" LINK="#FF6633"
ALINK="#FF0033"><BASEFONT SIZE="4">
<OL>
<LI><A HREF="http://www.facstaff.bucknell.edu/
rbeard/diction.html"> Web of Online
Dictionaries.</A> A multilingual dictionary site
that allows you to look up words in several
different languages through links to other
online dictionaries.
<LI><A HREF="http://www.notam.uio.no/~hcholm/
altlang/"> Dictionary of Alternative
Languages.</A> A multilingual slang dictionary
compiled collaboratively by Internet users.
<LI><A HREF="http://www.spellweb.com/">
SpellWeb.</A> If what you want is a correct
spelling, this site will compare two alternative
spellings and tell you which one is more popular
(according to Internet users).
<LI><A HREF="http://www.pcwebopaedia.com/">
Webopaedia.</A> Dictionary of computer
terminology with extensive links for each term.
</OL>
<A HREF="mailto:mbatsche@lonestar.utsa.edu">Send
me an e-mail.</A>
```

```
</BODY>
</HTML>
```

Figure 7.6 shows what it looks like in a browser.

Relative Links

The other common type of link is the abbreviated relative link. With relative links, because you're linking within the same directory, you don't need to specify the protocol, server, or complete path. The HTML uses the same container tags, so a relative tag would look like this: `<A HREF="index.html"> Back to Home page</A>`. That code tells the browser to look for the file `index.html` in the same directory as the file it's currently showing. Occasionally, you'll need to add another subdirectory name to the URL, particularly if you have more than one subdirectory in a larger directory. So, for example, to link a page to a page called `list` in another subdirectory called `stylebooks`, the link tag would read `<A HREF="/stylebooks/list.html">List of Stylebooks</A>`. This code tells the browser to look for another subdirectory called `stylebooks` in the same large directory containing the page currently being shown and to find the page called `list.html`. To go the other way, that is, to move up one level to the next level of directories in a site, use `../`. So, for example, to link a page called `list` in a directory called `stylebooks` that contained the current subdirectory, write `<A HREF="../stylebooks/list.html">List of stylebooks</A>`.

The code for relative links looks like this:

```
<HTML>
<HEAD>
        <TITLE>Links</TITLE>
</HEAD>
```

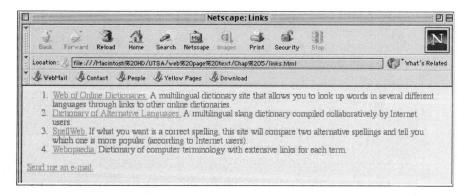

FIGURE 7.6 E-Mail Address

```
<BODY TEXT="#0033FF" VLINK="#33CC33"
BGCOLOR="#FFFF00" LINK="#FF6633"
ALINK="#FF0033"><BASEFONT SIZE="4">
<UL TYPE="square">
<LI><A HREF="welcome.html"><I>Home Page</I></A>
<LI><A HREF="assign1.html"><I>First
Assignment</I></A>
<LI><A HREF="journal.html"><I>Journal
Assignment</I></A>
<LI><A HREF="persnarr"><I>Personal
Narrative</I></A>
<LI><A HREF="prop.html"><I>Proposal</I></A>
<LI><A HREF="schedule.html"><I>Class
Schedule</I></A>
</UL>

</BODY>
</HTML>
```

Figure 7.7 shows what it looks like in a browser (italic text is used for variety here).

There's another good reason to use relative links: when you use the absolute URL, with protocol and domain name, the browser can't *cache* the page. That means that the page can't be placed in the browser's memory and it has to be called up fresh each time it's opened. This slows down your Web site by quite a bit, something many users find annoying. To keep your pages as fast-loading as possible, use relative URLs for pages on the same site.

Creating Relative Links

1. Put the cursor on your HTML page where you want the link to appear.

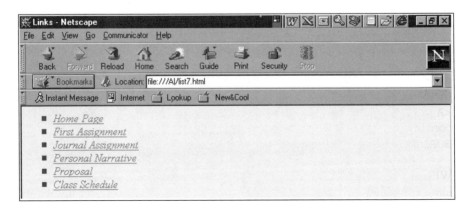

FIGURE 7.7 Relative Links

2. Type `<A HREF="url. address">`, filling in the relative address of the page on your site you want to link to.
3. Type the clickable text that will be underlined or otherwise highlighted so that the user can click on it and initiate the link.
4. Type `</A>`

■ TRY IT

Write the HTML and text for two pages, then create relative links to link the two. Remember to include the basic template on both pages: `<HTML> </HTML>`, `<HEAD></HEAD>`, `<TITLE></TITLE>`, `<BODY></BODY>`

If you've created multiple pages for either of the Web site assignments, you can link those pages here.

Using Graphics for Links

You can also use pictures as your clickable objects; for example, it's quite common to use buttons of some kind for relative links. The HTML for graphic links is the same as that used for text links; the only difference is that you put a graphic tag instead of text. The result looks like this:

```
<A HREF="http://www.cofah.utsa.edu/ecpc/mbatch/
default.html"><IMG SRC="images/house.gif"></A>
```

When text becomes linked, browsers will underline it to indicate that it's clickable. When a graphic is linked, it will be outlined with a square border in the link color. Sometimes this isn't a problem—if the color works with your graphic and if the border doesn't seem obtrusive. Figure 7.8 illustrates what the link border looks like.

However, there are times when you may want to turn off the border. In Figure 7.8, for example, you might prefer to have the graphic floating against the background without the square around it. To turn off the borders on graphics, use the same attribute and value you used with tables: `BORDER=0`. However, put them in the `IMG SRC` tag. So, for example, the tag for the graphic in Figure 7.8 is `<A HREF="list.html"><IMG SRC="seal.gif"></A>`. To get rid of that border, you'd write it as `<A HREF="list.html"><IMG SRC="seal.gif" BORDER=0></A>`. The graphic would then look like Figure 7.9.

Users can still tell that the graphic is a link by passing the cursor over it. When the cursor passes across a link in most browsers, it becomes a pointing hand (ready to click the link). However, if you're concerned that your users may not realize that the graphic is also a link (if, for example, the graphic is in a corner of the screen), leave the border on.

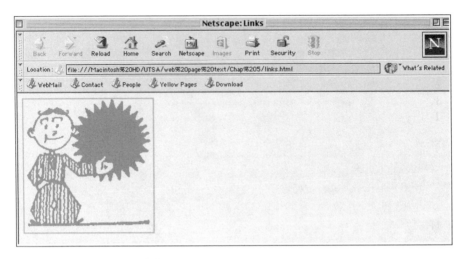

FIGURE 7.8   Graphic Link

Sometimes you may want text to accompany your button, particularly if the button itself doesn't indicate what's being linked. If you want the text to be linked along with the button, type the text after the `<IMG SRC= "image. gif">` but before the `</A>`. If you want the button to be linked but not the text, type the text after the `</A>`. In other words, linked text would look like this: `<A HREF="home.html"><IMG SRC="house.gif">Return to home</A>`. Nonlinked text would look like this: `<A HREF="home.html"> <IMG SRC="house.gif"></A>Return to home.`

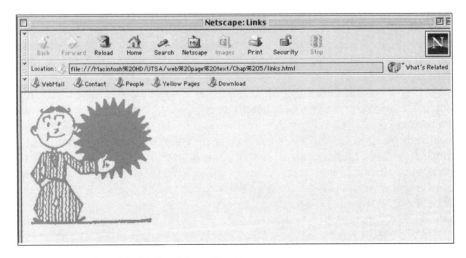

FIGURE 7.9   Graphic Link without Border

Creating Linked Graphics

1. Type `<A HREF="url. address">` (url. address will be the address of the page to which the user will jump by clicking the image.)
2. Type `<IMG SRC="image">` (Substitute the name and type of the graphic.)
3. Type `</A>` to complete the definition of the link.
4. If you want linked text along with the button, type the text after the `<IMG SRC= "image">` but before the `</A>`.
5. To remove the square border around the image, type `<IMG SRC="image" BORDER=0>`

■ TRY IT

Use the two pages you worked with in the previous exercise. Use a graphic for one of the links on one of the pages.

Anchor Links

The final type of link is called an anchor (sometimes called a *fragment*), and you use it when you need to link information on a single page. For example, say you were writing a page on soccer rules, beginning with the basic rules and working up to the more complex situations. Some of your more experienced readers might want to skip to the more complex information immediately. By providing a menu of topics at the top of the page, you allow them to skip down the page to the sections of interest, without having to scroll through everything in between.

Absolute and relative links both refer to URLs of other pages, both off-site and on-site. However, because anchors link to material on the same page, you don't need a URL. Instead, you create an anchor, a line of code, on the page and then refer to that anchor in your link. The HTML tag for your anchor is `A Name="anchor name"`. The anchor name can be anything, but for the sake of simplicity, it's usually easiest simply to insert the anchors next to any headings you use, and then to use the same wording as the heading for the anchor name. Because it's inside the angle brackets, the anchor name is invisible, and it can simply repeat the heading word. So, for a heading such as "What Is Offsides?" you'd make the anchor `<A Name="offsides">`. Technically, if your anchor name is only one word, the quotation marks are optional; however, to be on the safe side, include them.

Creating an Anchor

1. Place the cursor at the point on the page you want the user to jump to.

2. Type <A NAME="anchor name">. The anchor name will be the text you'll use to refer to the location on the page in your anchor link (e.g., a heading, a phrase in the text, an image, etc.).

■ ANCHOR NAME TIPS ■

There are a few things to keep in mind as you create your anchors:

- Insert anchor names as you write the HTML for your page, attaching them to section titles or other prominent points on the page.
- Place the anchor directly *before* the text you want to link to; that way that text will appear on the screen as soon as the user clicks the link.
- If the anchor name includes more than one word, you must put it in quotation marks. (It's a good idea to do this with one-word anchor names as well).
- Break up a long Web page by putting anchors on each section and referring to them in a table of contents at the top and bottom of the page.

Now you need to link to the anchor you've created. You can either link to an anchor on the same (long) page or to an anchor on another page of your site. The default place for a relative or absolute link to open is the top of the page; if you want to link to another part of a page on your site, use an anchor. You can even link to anchors on someone else's Web site, providing you know what the anchor names on that site are. (You can find them by looking at View Source Code or View Code in your browser.) Anchor links use the same basic container tags as the other types of links.

Linking to an Anchor on the Same Page
1. At the point you want to link to the anchor, type <A HREF="#
2. Type the anchor name that you created (result should be <A HREF= "#anchor name">).
3. Type the clickable text that users will click on to go to the anchor.
4. Type </A> to complete the link.

Linking to an Anchor on Another Page
1. At the point you want to link to the anchor, type <A HREF="url. address of page#
2. Type the anchor name that you created. (Result should be <A HREF= "url.address of page#anchor name">)

3. Type the clickable text that users will click on to go to the anchor.

4. Type </A> to complete the link.

Note: There should be no space between the address and #.

Here's a longer page with several anchors, including an anchor link to a second page (`compteam.html`). Notice the inclusion of `Return to Top` anchors so that readers can go back to the menu of anchors if they want to make another selection.

```
<HTML>
<HEAD>
        <TITLE>Links</TITLE>
</HEAD>
<BODY TEXT="#0033FF" VLINK="#33CC33"
BGCOLOR="#FFFF00" LINK="#FF6633"
ALINK="#FF0033"><BASEFONT SIZE="4">
<A Name="top">
<H2>League Soccer Rules</H2>
If you've played football or baseball, you may
find soccer rules somewhat simpler to understand
and follow. However, like any new activity,
you'll need to take some time to familiarize
yourself with the way the game is played. Younger
children, frequently age five to ten, may play by
modified rules. But here are some of the more
common soccer situations.
<UL>
        <LI><A HREF="#substitutes">Substitutes</A>
        <LI><A HREF="#fieldsize">Field size</A>
        <LI><A HREF="#cornerkicks">Corner Kicks</A>
        <LI><A HREF="#goalies">Goalies</A>
</UL>
The emphasis of the league is on noncompetitive
play for youngsters from 5-9; the intent of the
play is to develop individual soccer skills and a
positive attitude toward playing, as well as good
sportsmanship. We want children to enjoy
themselves first of all, and to learn about
competitive sports as a secondary goal.
<P>Older players may be involved in
<A HREF="compteam.html#tryouts">competitive
teams</A> selected on the basis of talent and
```

commitment. Tryouts for these competitive teams are held in the Spring and the Fall. Older players who prefer recreational play can elect to join a recreational team as an alternative to competitive play.
```
<A NAME="substitutes">
<H2>Substitutes</H2>
```
Players can be substituted after a team has thrown in, after a goal (scored by either team), after a goal kick (by either team), and if a player is injured. `<A HREF="#top">Return to Top</A>`
```
<A NAME="fieldsize">
<H2>Field Size</H2>
```
Players twelve and over play on regular-sized fields. Younger players use smaller fields, depending on the age group.`<A HREF="#top">Return to Top</A>`
```
<A NAME="cornerkicks">
<H2>Corner Kicks</H2>
```
Corner kicks restart the game if the ball crosses the goal line without scoring a goal and if it is touched by a defender before it crosses the goal. There are some rules associated with corner kicks.
```
<OL>
        <LI>The ball must be placed inside the
corner arc.
        <LI>The kicker can't touch the ball again
(after it has been kicked) until it is touched by
another player.
        <LI>Opponents must be 10 yards away from
the ball (3-5 yards for younger players).
</OL>
<A HREF="#top">Return to Top</A>
<A NAME="goalies">
<H2>Goalies</H2>
```
Goalies are allowed to use their hands only within the marked goal box (goalies are not used with players younger than six for safety reasons).`<A HREF="#top">Return to Top</A>`
```
</BODY>
</HTML>
```

Figure 7.10 shows what the page looks like in a browser. (You'll be able to see only the top half.)

■ TRY IT

Write the HTML and text for a page including at least one anchor link. (*Note:* This time, you'll need to include enough text to make the page scroll so that the anchor link can move down the page.) If you've written some longer pages for your text-based site, try using anchor links to break them up.

LINK RHETORIC

We've now covered the how part of linking: you've seen the most common types of links you can use on your Web pages. But we haven't really talked about the why: what you should link to and how you should integrate your links with your writing and design.

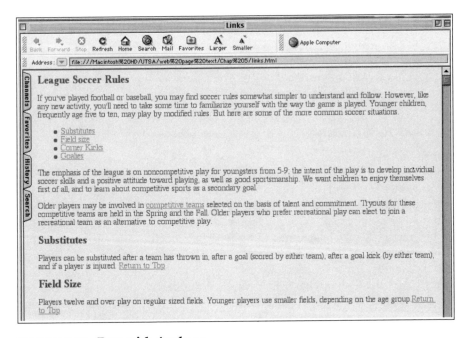

FIGURE 7.10 Page with Anchors

Links as Distractors

It's a rule of thumb that links should stand out: if readers can't recognize a link as a link, they won't click on it. But the very fact that links stand out means that they will contrast starkly with the surrounding text. (Jeffrey Veen, of *Hot-Wired,* the online magazine, refers to links as "little blue scars.") Is that bad? Not necessarily. On the other hand, a page of text that's peppered with links in a contrasting color may be hard to read, and readers may be tempted to click away rather than finish reading.

That may be the biggest problem with pages of links: they invite readers to leave rather than to stay for your content. One way around this impulse is to group your links together at the ends of pages or sections rather than sprinkling them through the text. Or put them in separate columns beside your text. You can even put all your links together on a separate page. All of these techniques avoid links that distract readers from what you have to say on your page. However, separating your links this way may mean that you're not using the hypertext aspect of the Web fully. The idea behind hypertext is that users will be able to find more information about a topic immediately, as soon as they feel a desire to know more. In general, you'll want to find a medium between extremes—not so many links that users feel distracted but enough of them to satisfy immediate curiosity.

When you create an in-text link, make sure that the word you use as the link is a clear indicator of where the link will take the reader. With a link such as "The goalie is the only player allowed to use hands inside the goal box," you'd expect a link to a description of the goalie's duties or position on the field. If the link were "The goalie is the only player allowed to use hands inside the goal box," you'd expect a description of what constitutes the goal box or perhaps a description of the soccer field.

Links as Reinforcers

You should also give careful thought to what pages you want to link to. A tendency among novice Web writers is to link to everything, calling the result "my favorite sites" or "cool pages." Although there may be some charm in eclectic site lists (you never know what you might find), they can also seem irrelevant. What do they have to do with the page you're currently on? The Web is a vast phantasmagoria of a place, full of strange, wonderful, and weird sites; but many surfers want links to be worth the trouble and bandwidth. What it boils down to is this: If your readers are going to click on a link, it should be a link that's clearly related to what's happening on the site. If your readers are interested enough in the topic to come to the site in the first place, they might like more information at related sites. At any rate, the readers are more likely to be interested in related sites than in sites on unrelated topics that appeal more to the site designer.

Used well, links can reinforce your information by supplying additional connections. Once you've caught your readers' interest in your content, help them to search further for more information.

Straightforward Linking

Finally, you need to give your readers some idea of where they're heading when they click on one of your links. For most readers, URLs won't mean much—unless someone already knows what a site is, the URL won't give much useful information. In fact, even the name of the site itself won't mean much unless the reader has already visited the site. (In which case, why click on the link?) If you're going to list sites, include a short description of what the site contains, so that the readers can decide whether or not to make the journey across the Web.

The earlier example of absolute links included short descriptions of all the sites that there were links for. For Webopaedia, the description was "dictionary of computer terminology with extensive links for each term." That's brief, but it gives the reader some idea of what's waiting at the other end of the link; then readers can decide whether the link is worth visiting.

■ WRITING ASSIGNMENT

Create a Web site that includes the types of links discussed in this chapter: an absolute link, a link to your e-mail address, a relative link, and an anchor link. Your site should also include the following:

- At least *two* linked pages
- At least *one* linked graphic, with or without a border
- Effective use of link rhetoric

Tips. Again, if you're working on your text-based site or your image-based site, you can use this assignment to create links between some of the pages. Make sure that every page has a link to another page, at least a link back to "home." (For more on site organization and navigation, see Chapter 8.)

TAGS USED IN CHAPTER 7

OPENING TAG	CLOSING TAG	EFFECT
<OL>	</OL>	Marks the beginning and end of an ordered list

OPENING TAG	CLOSING TAG	EFFECT
`<LI>`	None	Indicates list items
`<UL>`	`</UL>`	Marks the beginning and end of an unordered list
`<UL or LI TYPE="circle",` `"square", or "disc">`	`</UL>`	Changes the unordered list symbol to a circle, square, or bullet (the default symbol)
`<DL>`	`</DL>`	Marks the beginning and end of a definition list
`<DT>`	None	Indicates a term to be defined
`<DD>`	None	Indicates a definition
`<A HREF="url"*>`	`</A>`	Indicates a link
`<BODY LINK="x"**>`	`</BODY>`	Changes the default color of linked text or border around a linked graphic
`<BODY VLINK="x"**>`	`</BODY>`	Changes the default color of linked text that has been visited or border around a linked graphic that has been visited
`<A HREF="mailto:e-mail address">`	`</A>`	Indicates a link to an e-mail address
`<A HREF="url"><IMG SRC="image.gif"***>`	`</A>`	Indicates a linked graphic
`<A NAME="anchor name"†>`	None	Inserts an anchor name at a particular location on a Web page
`<A HREF="#anchor name"†>`	`</A>`	Indicates a link to an anchor on the same Web page
`<A HREF="url*#anchor name"†>`	`</A>`	Indicates a link to an anchor on another Web page

*Insert an absolute or relative URL.
**Insert a hexadecimal color number or one of the sixteen predefined colors.
***Insert the name and type of the graphic file.
†Insert a name for the anchor.

WEB SITES

http://bignosebird.com/navig.shtml
   Big Nose Bird's tutorial on site navigation; good advice and examples

http://info.med.yale.edu/caim/manual/interface/navigation.html
   Advice on navigation design from the *Yale Style Guide*

http://info.med.yale.edu/caim/manual/interface/navigation2.html
  More navigation and link advice from the *Yale Style Guide*

http://www.htmlgoodies.earthweb.com/tutors/sitelinks.html
  Page from Web designer Joe Burns about setting up links

http://www.useit.com/alertbox/9703a.html
  "The Need for Speed," article by information architect Jakob Nielsen on organizing and linking your site for speed

CHAPTER 8

HOW SHOULD IT WORK
Site Structure

BUILDING YOUR SITE

Because links allow you to approach information from several directions, there's sometimes a tendency to ignore the way your Web site is organized: just put those links in and let the reader figure it out, right? Unfortunately, that doesn't really work. Readers need help getting around your site, a sense of how the pages are related to each other, and even in some cases a defined path so they can see what page is related to what. Deciding how to organize and sequence your pages can be challenging, particularly if you include several topics and pages. In this chapter we'll discuss ways of ordering your material and making that order clear to your readers.

SITE ARCHITECTURE

Your first step will be to decide what pages you need to write to cover your subject. As we said in Chapter 4, you don't want long pages—it's better to keep the length to two screens or less if possible. That may mean subdividing your content into several linked pages, and then adding more pages as you need them. However, once you've decided how to divide up your content, how do you order and connect the pages?

Splash Page

The first page of your site has great importance: in many ways it will determine whether your readers stick around to see more or move on to another site. Some designers argue for a splash page, an opening page that sets the mood for the rest of your site. These pages usually include a graphic and a banner text of some kind: they're the ultimate attention getters that allow you to make a splash as soon as a user arrives at your site. (An example of a splash page for a guide to sushi in Figure 8.1. Notice the effective use of a fairly simple graphic to convey the idea and the style behind the site.)

If you choose to use a splash page, you can use a **META** tag to move your readers directly into your site (see the box Setting Up a Slide Show). However, you can also have something for users to click on if you don't want to make them wait.

■ SETTING UP A SLIDE SHOW ■

If you have a splash page on your site, you can move your visitor automatically from the splash page to the next page (your home page) by using the **META** tag. The **META** tag used in this way will display your page for a set number of seconds (you decide how long) and then dissolve into the next page. Before you set the number of seconds to display the splash page, be sure to measure how long it takes to load the initial graphic: you don't want the page to be dissolved before your readers have a chance to see it!

Using the META Tag to Show Your Splash Page

1. After the <HEAD> tag and before the </HEAD> tag, type

    <META HTTP-EQUIV="Refresh"

2. Type CONTENT="n For n substitute the number of seconds you want the splash page to be displayed.

FIGURE 8.1   Splash Page

3. Type `URL=` and add the relative URL for the next page.

4. Type `">`

The complete tag will look like this:

```
<META HTTP-EQUIV="Refresh" CONTENT="5 URL=
welcome.html">
```

If you want to add an option for clicking (for impatient visitors), link your opening graphic to your next page. (Be sure to include `BORDER=0` in the `IMG SRC` tag.)

Splash pages aren't always necessary; sometimes your visitors might be impatient at the delay. (For example, if they're trying to get some specific information and don't like waiting.) Moreover, if visitors will return repeatedly to your site—for example, if you have content such as a calendar that they'll need to consult over and over again—they may rapidly tire of having to click through a splash page to get to your information. In that case, you can begin your site with your home page and its menu of your site content. Assuming that your home page is well designed, you can create a good impression there too.

■ TRY IT

Do some surfing on the Web and look for particularly striking splash pages. Bring the URLs of the best two splash pages you find with you to class.

Home Page

Your home page should provide an introduction and overview of your site. Web designer Charlie Morris suggests that an effective home page will

- Provide an overview of the site and what's available on it.
- Project the right image for the site and the person and/or organization behind it.
- Make the identity of the person and/or organization behind the site obvious immediately.
- Set up a design that will be carried through in the other pages of the site, establishing certain repeating elements that will make the site seem unified (e.g., a logo, banner, navigation icons, etc.).
- Serve chiefly as a list of links to other pages on the site, although it may include a small amount of content: home pages can serve as tables of contents for the site as a whole.

Figure 8.2 shows the splash page for an Australian site on Aborigine dream stories; Figure 8.3 shows the home page that follows that opening splash page.

Notice the way in which the home page in Figure 8.3 picks up design elements from Figure 8.2, particularly the Stories of the Dreaming logo, which is

FIGURE 8.2   Splash Page

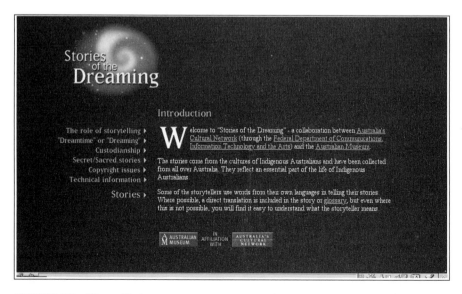

FIGURE 8.3   Home Page

used throughout the site. The home page effectively sets up the mood of the site and establishes a continuing design, but it also provides an overview of the site's organization through the menu–navigation bar on the left and a clear identification of the group–organization behind it in the buttons at the bottom of the page. It's a very effective beginning, both establishing the style of the site and allowing the reader to approach its content in a variety of ways.

■ TRY IT

Go back to the sites whose splash pages impressed you. Do they also have home pages? What design elements unify the splash page and the home page?

Site Organization

The rest of your site after the home page can be organized in a variety of ways. The simplest organization is linear; this means one page following another, with links forward and backward (see Figure 8.4).

Linear organization can work in some situations: if you have a set of pages that should be viewed in order, then a linear sequence may be your best choice. However, linear organization doesn't really take advantage of all that hypertext has to offer in terms of linking. Moreover, it doesn't allow your readers much control over the way they move through your information; thus they may not be happy with the arrangement.

More frequently, your site will be organized hierarchically. In other words, your site will resemble a pyramid with the home page at the top point and layers of pages branching off from it. Then if there are other pages related to those layers of pages, they'll branch into lower layers, and so on (see Figure 8.5).

Your site structure may ultimately look a little like an organization chart, with the home page at the top, but you can also think of it as an outline: a series of major topics subdivided into supporting topics, which may then be further subdivided and so on. However, most of these pages (at least the top two levels of the outline) need to have links on your home page, and the pages themselves should provide a return to home.

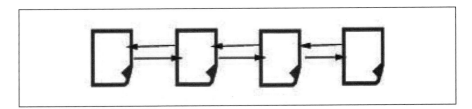

FIGURE 8.4   Linear Organization

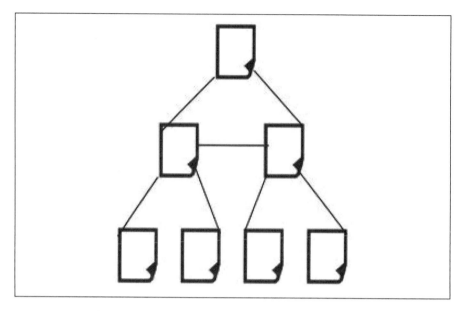

FIGURE 8.5   Hierarchical Organization

Another possibility for site organization is the network. In a networked site, all of the pages are interconnected, with no one page being central (see Figure 8.6).

This organization offers maximum flexibility for your users—they can go in any direction, pursuing any line of thought through your links. If you're putting together a creative site, allowing users to create their own organization of your material, this may well be the organization you'll want to use. However, although this organization scheme offers maximum possibilities for your readers, it's more time-consuming to construct because it requires more creative linking. The assumption behind network structure is that all the information on your site is equally important; thus you'll need to anticipate all the possible paths that your user might want to pursue. In some situations, your information may lend itself more readily to a hierarchical structure in which you give your readers a more definite indication of the way in which your information is related. It may be possible for you to use a modified network that more resembles a wheel, with your home page as the hub. That way all the pages branch off and return to a central point (see Figure 8.7).

Using a flowchart to diagram your site structure will help you see the way to arrange your site. Draw in the links between pages and consider how readers are likely to approach your information: what's the best path for them to take to find what they need when they need it? You can also use storyboarding, putting your pages onto 3-by-5 cards and then trying out various arrangements of them.

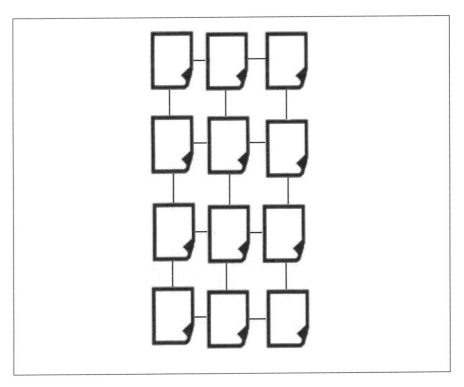

FIGURE 8.6   Network Structure

After you've decided on your site structure, your next concern becomes how to help your readers find their way around this structure.

■ TRY IT

Look at one of your favorite Web sites. How is it structured? Is there a site map that lays out the organization?

Navigation Bars

Navigation bars are a standard way to show off the organization of your site, so that readers can find what they want quickly. Typically, the navigation bar is a list of at least the major pages branching off from your home page. Sometimes this list includes the first level of subpages that branch off from your main pages as well; however, you may not want to go much deeper than that because you want the navigation bar itself to be relatively clear and simple to use. In Figure 8.8, a site for the Library of Congress's baseball card collection, the

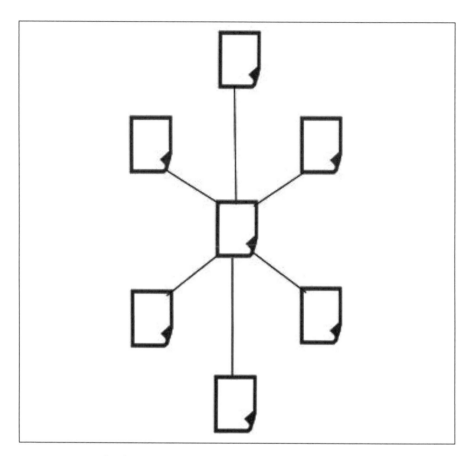

FIGURE 8.7   Wheel Structure

navigation bar is a simple horizontal list of words, beginning with Search by Keyword.

Your navigation bar can be words alone (as in Figure 8.8), or it can also be words and icons (as in Figure 8.9), or the navigation bar can use icons alone (as in Figure 8.10).

In Figure 8.10, the navigation bar consists of a series of colored rectangles across the top of the screen, with no indication of what they stand for. Icons alone are perhaps the riskiest choice, because the icons should usually be transparent; that is, there should be no confusion about what topics they represent. On the Alaska 360 site, the designers are banking on the attraction of the panoramic views of the Alaskan wilderness to make users click on the mysterious buttons to find out what they represent.

You can place your navigation bar down the side of your page or across the top or bottom. If you have a long page, you can even repeat the navigation bar at both the side or top and the bottom.

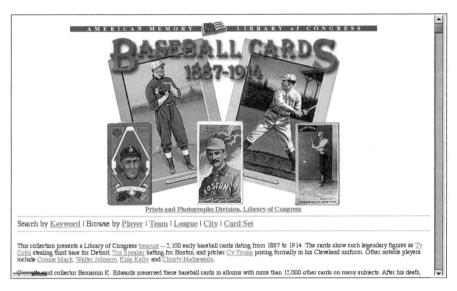

FIGURE 8.8   Word Navigation Bar

Library of Congress, Prints and Photographs Division

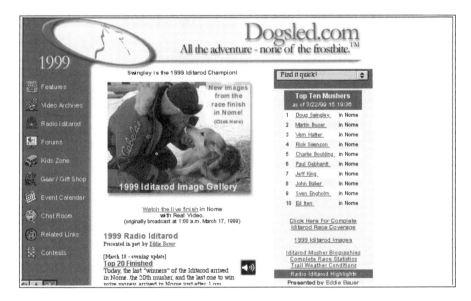

FIGURE 8.9   Word and Icon Navigation Bar

Word Bar.   The easiest way to create a navigation bar is simply to use words for your links. Then you can place the links in a table so that they run across the top of the page or down the side. For example, take a site with four pages

FIGURE 8.10   Icon Navigation Bar

beyond the home page: soccer schedule, youth soccer rules, red team tryouts, and frequently asked questions.

To run those links across the top or bottom of the page, you could create a one-row table with a link in each cell. If you wanted to add a little pizzazz, you'd make the cell backgrounds a different color from the page background. The HTML for a row of buttons across the bottom of the page would look like this:

```
<HTML>
<HEAD>
      <TITLE>Navigation Bar</TITLE>
</HEAD>
<BODY BGCOLOR="#FFFFFF" LINK="#000000" BASEFONT=
"4">
<H1>YSO Soccer</H1>
Welcome to the Youth Soccer Organization Soccer
site. Here we provide this year's schedule, a
brief overview of youth soccer rules,
information about tryouts for the competitive
red team, and some answers to questions. If you
need more information than you find on these
pages, please e-mail us at <A HREF="mailto:
yso@goodmail.com">yso@goodmail.com</A> Good luck
and good soccer!
```

```
<TABLE BORDER="0" CELLSPACING="10">
        <TR>
                <TD BGCOLOR="#0000FF" ALIGN=
"Middle"><FONT SIZE="4">
                <A HREF="schedule.html"><FONT
COLOR="#FFFFFF">Schedule</A>
 </TD>
                <TD BGCOLOR="#0000FF"ALIGN=
"Middle"><FONT SIZE="4">
                <A HREF="rules.html"><FONT COLOR=
"#FFFFFF">Rules</A>

 </TD>
 <TD BGCOLOR="#0000FF"ALIGN="Middle"><FONT SIZE=
"4">
                <A HREF="tryouts.html"><FONT
COLOR="#FFFFFF">Tryouts</A>

 </TD>
 <TD BGCOLOR="#0000FF"ALIGN="Middle"><FONT SIZE=
"4">
                <A HREF="questions.html"><FONT
COLOR="#FFFFFF">Frequently Asked Questions</A>

 </TD>
        </TR>
</TABLE>
</BODY>
</HTML>
```

Figure 8.11 shows what it would look like in a browser.

It looks okay, but not particularly elegant. Moreover, a horizontal navigation bar across the bottom of the page might run the risk of getting lost if the

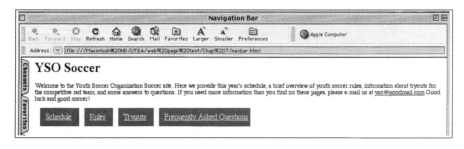

FIGURE 8.11 Horizontal Navigation Bar

reader had a small browser window that cut off the bottom of the page. You could get around that by putting the buttons across the top of the page, but then the page would be unbalanced. Let's try putting those buttons down the side. This gets a little more complex because now you have to put both the navigation bar and the text into the table so that they'll line up side by side. Put the text into a single column that will use a ROWSPAN to spread across the four button rows. If you add a WIDTH attribute to both the table and the cell that contains the text, you can make the text lines more narrow and push them to the top of the table. The HTML looks like this:

```
<HTML>
<HEAD>
      <TITLE>Navigation Bar</TITLE>
</HEAD>
<BODY BGCOLOR="#FFFFFF" LINK="#000000" BASEFONT=
"4">
<TABLE BORDER="0" CELLSPACING="10"
CELLPADDING="10" WIDTH="500">

      <TR>
              <TD BGCOLOR="#0000FF" ALIGN=
"Middle"><FONT SIZE="4">
              <A HREF="schedule.html"><FONT
COLOR="#FFFFFF">Schedule</A>

  </TD>
              <TD BGCOLOR="#FFFFFF" ROWSPAN="4"
WIDTH="300"><H1>YSO Soccer</H1><FONT SIZE="4">
Welcome to the Youth Soccer Organization Soccer
site. Here we provide this year's schedule, a
brief overview of youth soccer rules,
information about tryouts for the competitive
red team, and some answers to questions. If you
need more information than you find on these
pages, please e-mail us at <A HREF="mailto:
yso@goodmail.com">yso@goodmail.com</A> Good luck
and good soccer!</FONT>

              </TD></TR>
              <TR><TD BGCOLOR="#0000FF"ALIGN=
"Middle"><FONT SIZE="4">
              <A HREF="rules.html"><FONT COLOR=
"#FFFFFF">Rules</A>
```

```
     </TD></TR>
     <TR><TD BGCOLOR="#0000FF"ALIGN="Middle"><FONT
SIZE="4">
                    <A HREF="tryouts.html"><FONT
COLOR="#FFFFFF">Tryouts</A>

     </TD></TR>
     <TR><TD BGCOLOR="#0000FF"ALIGN="Middle"><FONT
SIZE="4">
                    <A HREF="questions.html"><FONT
COLOR="#FFFFFF">Frequently Asked Questions</A>

     </TD>
          </TR>
     </TABLE>
     </BODY>
     </HTML>
```

In a browser, it would look like Figure 8.12.

If you wanted to have just a stacked list of words, rather than buttons, you could give all of the table cells the same background. The result would look like Figure 8.13.

Creating a Horizontal Button Navigation Bar

1. Create a one-row table, using `<TABLE BORDER=0><TR>`
2. In the first table cell, change the background color to make the cell distinct from the page; for example, `<TD BGCOLOR="#0000FF">`
3. Add any other effects such as text color or size.

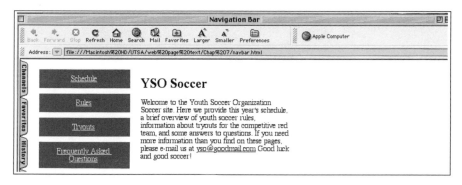

FIGURE 8.12 Vertical Navigation Bar

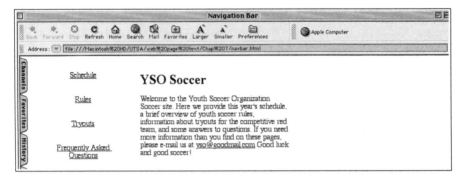

FIGURE 8.13   Navigation Bar without Buttons

4. Type the link, using clickable text.
5. Create another table cell in the same way for each additional link.
6. Type </TR></TABLE>

Creating a Vertical Text Navigation Bar

1. Create a table, using <TABLE BORDER=0><TR><TD>
2. In the first cell, type the first link on your navigation bar.
3. Create a second cell, using ROWSPAN to stretch the cell over the number of rows needed to accommodate the links on your navigation bar.
4. Enter text in the second cell.
5. Create cells for the remaining links on your navigation bar.
6. Close the table with </TD></TR></TABLE>

Note: You can also add formatting to the table; for example, change the font size and/or color, along with the table background.

■ TRY IT

Create the HTML and text for a page that includes either a horizontal or vertical navigation bar. Include at least three links. Remember to include the basic template: <HTML></HTML>, <HEAD></HEAD>, <TITLE></TITLE>, <BODY></BODY>

You can add navigation bars to either your text-based or image-based site, if they're appropriate to your design.

Icon Bar. Creating a navigation bar with icons can be done in one of two ways: you can use a series of linked graphics, or you can use what's called an image map, a large graphic with defined, linked areas. We'll cover only linked graphics here because they're easier to construct. Image maps can be useful, but it's easier to construct them using an authoring program such as Front Page or a Web-graphics program such as Image Ready rather than trying to write the HTML from scratch.

The downside of small linked graphics is the size of the file and its download time: each graphic will have to load separately. You can offset this liability somewhat if you keep the graphics quite small: no more than eight or ten kilobytes (K) at most.

You already know the HTML for creating a linked graphic:

```
<A HREF="questions.html"><IMG SRC="qmark.gif"
ALT="Question Mark" BORDER="0"></A>
```

For a navigation bar, you'll want to add a text caption to explain what the icon refers to. That would look like this:

```
<A HREF="questions.html"><IMG SRC="qmark.gif"
ALT="question mark" BORDER="0">Frequently Asked
Questions</A>
```

To make the text a little smaller (so it will fit with the icon), you'll add a FONT tag; you can also put a line break (BR) after the graphic so that the words are underneath the icon. All of that would look like this:

```
<A HREF="questions.html"><IMG SRC="qmark.gif"
ALT="question mark" BORDER="0"><BR><FONT SIZE=
"2">FAQs<FONT></A>
```

If you add icons for the other links on the navigation bar as well, the HTML will look like this:

```
<HTML>
<HEAD>
      <TITLE>Navigation Bar</TITLE>
</HEAD>
<BODY BGCOLOR="#FFFFFF" LINK="#000000" BASEFONT=
"4">
<TABLE BORDER="0" CELLSPACING="10"
CELLPADDING="10" WIDTH="500">
      <TR>
```

```
                <TD ALIGN="middle">
                <A HREF="schedule.html"><IMG
SRC="sched.gif" ALT="schedule"
BORDER="0"><BR><FONT SIZE="2">Schedule<FONT></A>

  </TD>
                <TD WIDTH="400"ROWSPAN="4">
<H1>YSO Soccer</H1><FONT SIZE="4">
Welcome to the Youth Soccer Organization Soccer
site. Here we provide this year's schedule, a
brief overview of youth soccer rules,
information about tryouts for the competitive
red team, and some answers to questions. If you
need more information than you find on these
pages, please e-mail us at <A HREF="mailto:
yso@goodmail.com">yso@goodmail.com</A> Good luck
and good soccer!</FONT>

                </TD></TR>
                <TR><TD ALIGN="middle">
                <A HREF="rules.html"><IMG
SRC="rules.gif" ALT="rules" BORDER="0"><BR><FONT
SIZE="2">Rules<FONT></A>
  </TD></TR>
  <TR><TD ALIGN="middle">
                <A HREF="tryout.html"><IMG SRC=
"try.gif" ALT="tryouts" BORDER="0"><BR><FONT
SIZE="2">Tryouts<FONT></A>
  </TD></TR>
  <TR><TD ALIGN="middle">
                <A HREF="questions.html"><IMG SRC=
"qmark.gif" ALT="questions" BORDER="0"><BR><FONT
SIZE="2">FAQs<FONT></A>
  </TD></TR>
</TABLE>
</BODY>
</HTML>
```

Figure 8.14 shows how it would look in a browser.

Once you've developed your navigation bar, you'll use the same one (or an abbreviated version) on other pages in your site as well. The navigation bar can become a way of unifying your site; it can be a common design feature from one page to the next. In addition, placing the navigation bar in the same place on every page makes it easier for readers to find: when they want to move to another page, they'll look in the same spot.

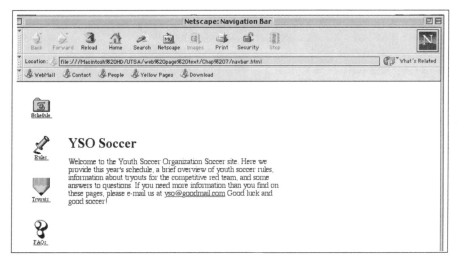

FIGURE 8.14   Icon Navigation Bar

Creating a Vertical Icon Navigation Bar

1. Create a table, using `<TABLE BORDER=0><TR><TD>`
2. In the first cell, type the first link on your navigation bar, using a combination of graphic and text (turn off the border by including `BORDER="0"` in the `IMG SRC` tag).
3. Create a second cell, using `ROWSPAN` to stretch the cell over the number of rows needed to accommodate the links on your navigation bar.
4. Enter text in the second cell.
5. Create cells for the remaining links on your navigation bar and insert graphic and text for each.
6. Close the table with `</TD></TR></TABLE>`

Note: You can change the size and position of the text accompanying the graphic links, using the `FONT` tag and `BR` after the graphic.

■ TRY IT

Create the HTML and text for a page with an icon navigation bar. Include at least two icons. Remember to include the basic template: `<HTML></HTML>`, `<HEAD></HEAD>`, `<TITLE></TITLE>`, `<BODY></BODY>`

You can add an icon navigation bar to your text-based or image-based site, if it would be appropriate to your design.

Page Bottom Navigation Bars

It's a good idea to place a text version of your navigation bar at the bottom of your pages (along with the formal version along the top or side), so that the reader doesn't have to scroll back to the top of the page to see the bar. This is particularly helpful if you're using graphics as part of your navigation bar. Text bars are not only a courtesy to the reader (in that they save scrolling); they can also make it easier for readers with text-only browsers or slow modems to view and use your site. On these bottom bars, the text links are separated by vertical lines. The code for a bottom bar is fairly simple:

```
<A HREF="schedule.html">Schedule</A>|
<A HREF="rules.html">Rules</A>|
<A HREF="tryout.html">Tryouts</A>|
<A HREF="questions.html">Questions</A>
```

Just type your links separated by vertical lines. On the page the result will look like this:

Schedule|Rules|Tryouts|Questions

Creating a Horizontal Text Navigation Bar

1. At the bottom of your page, type your first link, using clickable text.
2. Type a vertical line (e.g., |; frequently the vertical line is found on the same key as the forward slash).
3. Type your second link, using clickable text.
4. Type another vertical line.
5. Repeat with your other main links.

■ TRY IT

Use the page you created for either of the previous two exercises. Insert a horizontal text navigation bar at the bottom.

Dead-End Pages

You should keep one final point in mind as you put your pages together: every page you have on your site should have at least one link. Pages that have no links at all are sometimes called dead-end pages: they have no way out. It's true that your readers could return to the previous page by hitting the Back button on their browsers, but why make them do that? At the very least, all of your pages should link to your home page. If readers have come to a page because it offers supplemental information related to one particular page in your site, then you should provide a link that will return them to the page where they were originally. Don't make readers thread their way back through your site, trying to find a way out via their Back button. They're liable to leave your site and not come back!

◼ NAVIGATION BAR TIPS ◼

- Keep your navigation bars small, in terms of both file size and dimensions. Don't let the navigation bar take up too much memory or too much space on the screen.
- Repeat the navigation bar in text form at the end of the page; make allowances for text-only browsers.
- Keep the number of choices on the navigation bar limited; just present the top level of pages on your site. Don't overwhelm your reader with options.
- On the other hand, don't subdivide your pages so far that your readers have to go through too many pages to get where they want to go. The rule of thumb: don't make a reader travel through more than five pages to get from point A to point B.
- Avoid dead-end pages: every page should have at least one link.

◼ WRITING ASSIGNMENT

Create a Web site that is structured using one of the organization plans discussed in this chapter, that is, linear, hierarchical, network, or wheel. Each page should have a navigation bar. Your site should also include:

- At least *three* linked pages
- At least *two* paragraphs of text per page

Tips.   If you're working on one of the Web site assignments, you can use this assignment as an opportunity to work with the organization of your site. Make sure that the organizational structure you choose is appropriate to your content; that is, don't simply choose a linear organization because you're more familiar with that concept from paper documents. How are users likely to approach your information? What kinds of connections would they like to be able to make? How can you help them to get around your site quickly and easily so that they find the information they want and need?

Web Site 2 ■ An Image-Based Site

Your second Web site will be one whose purpose is predominantly to entertain and please your user visually. Your main content on this site will be images of some kind, although you'll probably have short text passages as well.

Jeffrey Veen is also the source of the "Web site as gallery" concept. Here's what he has to say in *HotWired Style*:

> ...the gallery offers a controlled presentation of its artwork. It is a carefully orchestrated space designed to give every visitor the same, meticulously curated aesthetic experience. (2)

You'll draw on the information in the first six chapters of the book here, presenting your text in a readable, appealing way, but also presenting images and colors that will please your reader visually. You'll probably need to draw on information from the last two chapters as well so that you can create links between the pages on the site and between this site and others. Your instructor will decide how much he or she wants you to do.

TIPS

1. Clearly you'll have an easier time on this assignment if your topic is visually oriented to begin with; logical choices include museum and gallery exhibits, artistic retrospectives, and performances. Another possibility might be a "guided tour" Web site for a location, either real or imaginary. You can transform a topic into a visual experience, such as a site that illustrates a concept or text through images. However, whatever the topic, images are your central focus here.

2. You'll need to concentrate on producing a visually unified and aesthetically pleasing site here; however, navigation will be an important factor because users may well skip around rather than proceeding linearly. Look at the advice for organizing your site in this chapter. Consider what navigation options to provide to your users so that they can move around more freely. Should every page link to a central "hub" page? Do you want hierarchies and "subsets"? How are people likely to approach this information and how can you make it easier for them?

3. You'll need a minimum of three pages, but this time you may want to include a "splash" or entry page.

4. You'll need to pay attention to image file size. Remember: as a rough rule of thumb, one kilobyte of information (i.e., 1K) requires an average of one second to download. If your file is larger than 60K, you're making your users wait for more than a minute with a largely blank screen. Bad idea! Work on getting GIFs down to as few colors as you can while still

retaining their quality and on saving JPEGs at the correct quality setting (not everything has to be "maximum").

5. Consider designing your site around a visual metaphor or concept: a gallery wall, for example, or a walking tour of a location. Using a visual metaphor gives you a central idea to unify your pages so that they all have the same look, a desirable quality on a site like this.

6. Layout here will be determined by your content: a "guided tour" site and an "exhibition" site will have different looks. However, you'll probably be working with single-screen pages here, rather than the two-and-a-half-screen scrolling pages you used on your text site. Look at the example sites above or in the Viewer Sites listed at the end of Chapter 4 for ideas that you can apply to your own design.

7. You'll want a coherent page design, as you had on the text-based site. Make sure that all of your pages use the same general template, so that users can be certain that they're still on the same site as they travel from page to page. Use the same navigation features throughout (see Chapter 8), and make sure they're in the same place on every page.

WEB SITES

Example Sites

http://memory.loc.gov/ammem/bbhtml/bbhome.html
Library of Congress baseball card collection

http://nmd.hyperisland.se/studentzone/crew2/martin_ragnevad/
Sushi Guide

http://www.dogsled.com/
Dogsled site: "all the fun; none of the frostbite"

http://www.dreamtime.net.au/
Stories of the Dreaming: Aborigine dream stories

http://www.360alaska.com/
360 Alaska: panoramic views of Alaska

Site Structure and Navigation

http://home.netscape.com/computing/webbuilding/studio/feature19980729–1.html
"Seven Deadly Sins of Information Design" by Drue Miller of vivid studios

http://home.netscape.com/computing/webbuilding/studio/feature19980827–2.html"
"Click Here: How Not to Write for the Web" by Drue Miller

http://info.med.yale.edu/caim/manual/sites/site_structure.html
   Yale Style Guide pages on structure

http://wdvl.internet.com/Location/Navigation/101/
   Tutorial on navigation from Charlie Morris

http://wdvl.internet.com/WebRef/Navigation/Design.html
   "Designing for Navigation" by David Walker

http://www.builder.com/Graphics/Design/ss1a.html
   Ben Benjamin's advice on site organization

http://www.zeldman.com/askdrweb/design.html
   Ask Dr. Web (Jeffrey Zeldman) advice on site design

http://www.zdnet.com/devhead/stories/articles/0,4413,2253058,00.html
   Jakob Nielsen's "Users First: How to Structure Your Website"

▪ ▪ ▪ ▪ ▪ ▬▬▬▬▬▬▬▬▬▬▬▬▬▬▬▬▬▬▬▬▬

CHAPTER 9

THE FUTURE

Cascading Style Sheets

HTML 4.0

The Web doesn't stand still, although some of us might wish that it did occasionally so that we could catch up! Along with new browser and software versions, new versions of HTML have been approved periodically by the World Wide Web Consortium. The newest of these versions (as this book is being written) is HTML 4.0, a version of HTML that will have far-reaching effects once it's fully implemented by browsers.

Cascading Style Sheets (CSS) were designed to provide the features that writers and designers had found lacking in the original versions of HTML. CSS allows you to choose fonts, to specify font sizes using familiar measurements such as points (rather than the HTML default 1–7), to establish margins and line spacing, and, best of all, to position elements on a page by specifying an exact location. That's the good news. And the bad?

Right now, only parts of CSS can be used in existing browsers. More CSS features are implemented with each new browser release, but as yet no existing browser will allow you to use all of the elements of CSS, and the "gaps" in implementation can be very frustrating! The World Wide Web Consortium is working with browser manufacturers to ensure that future Web browsers will be fully compliant with CSS. Moreover, at the time this book is being written (1999–2000), Netscape and Internet Explorer have both stated that their Version 5 upgrades will be fully compliant with CSS specifications. In the near future, browsers will exist that will implement all the capabilities of CSS.

What about right now? Several sites on the Web keep track of which browsers are capable of displaying which features of CSS (check the sites listed at the end of this chapter), and you can always check one of them to find out if a particular feature will work. But perhaps the easiest way to see whether your CSS will show up the way you want it to is to look at it in a browser. Thus we're back to the advice from Chapter 1: test, test, test! Some pages may show up well in one browser but not in another, or in one browser version but not in another. This is an aspect of the Web that's currently in considerable flux. Keep tuned, and keep trying!

171

CASCADING STYLE SHEETS

Perhaps the easiest way to think of CSS is to compare it to the style sheets in word processing programs. In most full-featured word processors and page-layout programs (such as Microsoft Word, WordPerfect, Quark Xpress, and Pagemaker), you can specify a certain combination of style features for each part of your document. So, for example, if you want your top-level headings to be twelve-point bold Helvetica, you can define that heading style in your style sheet. Then every time you want to create a new heading, you just click on the heading style and the text is automatically changed to twelve-point bold Helvetica. Moreover, if you decide later that you'd like those headings to be in fourteen-point Helvetica rather than twelve point, you can make the change in the style sheet, which will automatically change all the headings in your paper.

Clearly, style sheets are a great time-saver, but they also help to unify documents by making sure that every time you use a particular element (all your top-level headings, for example) that element will look the same. This can help your readers make sense of your information more quickly and also help them understand the organization of your page at a glance.

Cascading Style Sheets give you the same kind of ability to define your styles and to keep them uniform throughout your Web site. They also allow you to change the default HTML styles, such as the ones for headings. Rather than the bolded oversize style that HTML sets for H1, for example, you can choose another type font, size, treatment, and even color.

External and Internal Style Sheets

You can create style sheets in several different ways using CSS. External style sheets are created on a separate page of your Web site. You write all of your style definitions (rules) on that page, then save it as a text-only file with the extension `.css`. Then you write a line of code—`<LINK REL=stylesheet TYPE= "text/css" HREF="filename.css">`—in the `<HEAD>` section of your page.

External style sheets are supposed to work like the style sheets in word processors, but they aren't really reliable in all situations yet. For now, we'll focus on style sheets that you insert into the page you're writing.

■ HIDING YOUR STYLES FROM OLDER BROWSERS ■

As I said at the beginning of this chapter, only the most recent browser versions can read style sheet codes. Unfortunately, that means that some older browsers not only won't interpret your styles correctly but also may show your CSS code as if it were text. To avoid confusing older browsers, you can enclose your internal style sheets in the tags used for comments:

```
<!--<STYLE> Your CSS code</STYLE>-->.
```

This will keep the older browsers from showing the code, but it won't prevent newer browsers from interpreting the styles correctly. If you actually want to add comments as well as conceal your styles, begin the comment with /* and end it with */.

Your hidden text would look like this:

```
<HEAD>
        <TITLE>CSS Styles</TITLE>
<STYLE TYPE="text/css">
<!--
P{font-family:verdana,arial,helvetica,sans-
serif;}
-->
</STYLE>
</HEAD>
```

Internal Style Sheets

Internal style sheets are placed in the `<HEAD>` section of your page. They use a set of container tags: `<STYLE></STYLE>`.

Creating an Internal Style Sheet
1. After `<HEAD>` but before `</HEAD>`, type `<STYLE TYPE="text/css">`
2. Type your CSS styles.
3. Type `</STYLE>`

The styles you define between `<STYLE TYPE="text.css">` and `</STYLE>` will apply to all the elements on the page. So if you wanted a different type size for H1, you could define the new size in the `<STYLE>` section. Then every time you use `<H1>` on the page, it will use the size you defined in your `<STYLE>` section—and you won't have to keep adding `<FONT SIZE="n">` to each `<H1>`, as you did when you wanted to make such a change in previous versions of HTML.

You can also add CSS styles to individual HTML tags, just as you did with attributes such as `SIZE`. You just use `STYLE` as an attribute inside the tag. The result looks like this: `<H1 STYLE="font-size:12pt;color:blue">`. So if you wanted to make one `<H1>` heading red, eighteen-point Helvetica, you'd write `<H1 STYLE="font:18pt Helvetica;color:red">`.

Creating an In-Line Style
1. Inside the tag you want to change (e.g., `<P>`), write `STYLE="`

2. Write all the style rules you want to include and end with quotation marks and end bracket (e.g., `<P STYLE="font-family: Arial, Helvetica, sans-serif; font-size: 10pt">`).

3. Be sure to include an end tag (e.g., `</P>`) at the end of the section to turn off the style.

■ WHY CASCADING? ■

The cascade part of Cascading Style Sheets comes from the fact that several different style sheets can apply to a given Web page. However, even though you might have an external style sheet as well as an internal style sheet and tags where you've used the **STYLE** attribute on a particular page, all of these styles will cascade in the order in which they're applied from specific to general. In other words,

The **STYLE** attribute will override	External style sheets, internal style sheets, and default HTML style
Internal style sheets will override	External style sheets and default HTML style
External style sheets will override	Default HTML style

Thus you could define **H1** in two or three separate ways—one on an external style sheet, another on an internal style sheet, and another using the **STYLE** attribute in the **H1** tag—but all of them would override the default HTML definition of what **H1** looks like.

CSS BASICS

Now you know where to define the CSS styles, but what goes into those styles themselves? For this you need a new set of terms.

A CSS *rule* defines how HTML elements will look and behave when they're viewed through a browser. All rules, whether external, internal, or within an HTML tag, have at least two parts: a selector and a property.

- The *selector* is a word or a set of characters that identifies the style that's being defined. It can be an HTML tag (e.g., **P**) or a *class* or *ID* selector. We'll talk more about classes and IDs below.
- The *property* is what's being defined. Properties include the way an element looks (e.g., color, size) and the way it acts (e.g., position).

- The *value* means the same thing here as it did with the HTML attributes: it's the definition of the property. Let's say you have this property definition: `color: green`. In this definition, `color` is the *property,* whereas `green` is the *value.*

The format for CSS rules is the same wherever you place them.

```
Selector{property:value; property:value;}
```

Writing a CSS Rule

1. Write the selector followed by a bracket (e.g., `P{`).
2. Write the property, followed by a colon, and the value; if you have more than one property, separate them with semicolons (e.g., `P{font-size:10pt;color:red;`).
3. Close with a final bracket (e.g., `P{font-size:10pt;color:red;}`).

Although you don't have to have a semicolon after the final value, using one will make sure that you won't forget to add one if you add another property later.

Obviously, not all values will work for all properties: `font-size` could never be defined as `brown`. If you're ever confused about whether a property and value will work together, or whether a particular measurement (e.g., inches) will work with a certain property, you can consult one of the reference sites listed at the end of this chapter.

Classes and IDs

Sometimes you may want to create special styles that you'll apply to some elements on a page but not to others. Say you wanted to emphasize some paragraphs but not all of them. You wouldn't want to use `P` as the selector when you defined the style because that would apply the style to all of the paragraphs on the page. Instead you can create a special type of selector called a `CLASS`, which you can apply to the paragraphs you want to emphasize.

Creating a Class Selector

1. After `<STYLE>` but before `</STYLE>`, write the selector name, preceded by a period and followed by a bracket (e.g., `.cool{`).
2. Write the properties and values that define the class; separate with semicolons and follow with a closing bracket (e.g., `.cool{color: blue;}`).

You can use classes with any HTML tag, as long as the properties in the class work with that tag. (For example, you couldn't use the cool class created in the preceding example with IMG SRC because you can't change the color of an image with HTML.) To use a class with an HTML tag, you insert it inside the tag.

Using a Class with an HTML Tag

1. After you define the class in your STYLE section, type the tag you want to associate with the class on your page (e.g., <H2).
2. Type CLASS="classname">, adding the name of the class you created (e.g., <H2 CLASS="cool">).
3. Add the close tag after filling in the text (e.g., </H2>).

You can apply the same class to several different tags on the same page—it's not limited to one particular element or type of element. So, for example, I could apply my cool class to paragraphs or headings or even to table cells—wherever I wanted the type to look cool.

Classes give you a very handy way to define different styles and apply them only when you need them. Incidentally, you can call your classes whatever you want, but keep the names short and don't use more than one word. (You also can't have spaces in the name.)

IDs function the same way that classes do, plus they give a selector a unique name, which is sometimes necessary for writing scripting languages such as JavaScript. They're also handy when you use CSS positioning rules. Unlike classes, you can use an ID only once on a page, because it's supposed to give an element a unique name. If you're not using either a scripting language or CSS positioning rules, it's easier to use classes.

Creating an ID Selector

1. After <STYLE> but before </STYLE>, write the selector name, preceded by a number sign (#) and followed by a bracket (e.g., #hot{).
2. Write the properties and values that define the ID; separate them with semi-colons and follow with a closing bracket (e.g., #hot{color:orange;}).

Using an ID with an HTML Tag

1. Type the tag you want to associate with the ID (e.g., <H3>).
2. Type ID="IDname", adding the name of the ID you created (e.g., <H3 ID="hot">).
3. Add the close tag after filling in text (e.g., </H3>).

DIVs and SPANs

Classes and IDs allow you, in a sense, to create your own HTML categories; DIV and SPAN are tags that allow you to create sections of your Web pages where you can apply your CSS.

The DIV tag creates a division on the page, separated by a break above and below, allowing you to set off a page section and apply a particular style.

Using DIV to Create a Division

1. At the beginning of the section, type <DIV
2. Add CLASS="classname">, adding the name of the class you want used in the section *or*
3. At the beginning of the section, type <DIV STYLE="properties">, adding the properties you want used in the section.
4. Type the text and tags for the section.
5. Type </DIV> to end the section.

The SPAN tag creates a section that has no break from what went before. Thus you can use SPAN to add style features within paragraphs or even within sentences; for example, if you wanted to add a yellow background, like a highlighter, to a single line of text for emphasis, you could use SPAN to do that. You could also utilize SPAN to create a drop cap: you can use it with sections as small as a single letter.

Using SPAN to Create a Section

1. At the beginning of the section, type <SPAN
2. Add CLASS="classname">, adding the name of the class you want used in the section *or*
3. At the beginning of the section, type <SPAN STYLE="properties">, adding the properties you want used in the section.
4. Type the text and tags for the section.
5. Type </SPAN> to end the section.

CSS TEXT RULES

Now that you've seen the basics of how to set up CSS on your page with internal style sheets or within HTML tags, let's look at some of the properties you can use. The CSS properties for text are probably the most widely implemented properties in current browsers. Some of these properties will even show up in version 3 browsers, and many more are available in versions 4 and 5.

Font-Family

The `font-family` property allows you to select a particular font for your text; however, as usual, your page can't be displayed using a font that isn't loaded on a reader's computer. CSS gets around this problem by allowing you to select more than one font and then list them in order of preference. So your rule might look like this:

```
H2{font-family:verdana,arial,helvetica,sans-
serif;}
```

The browser will go down your list in order until it finds a font loaded on the reader's computer that matches one of the names; in this example, if the browser doesn't find Verdana on the reader's computer, it will look next for Arial, then Helvetica, and then for a generic sans serif font (see Type Terms box in Chapter 6). You should end your list with a generic choice, as in this list: serif, sans-serif, cursive (script fonts such as Zapf Chancery), fantasy (display fonts, see Glossary), or monospace (e.g., Courier). That way the browser will try to find a font that looks roughly like the font you originally specified. (*Note:* Cursive and fantasy may give you unexpected results because cursive and display fonts differ radically from font to font and the browser will choose the first one it comes to on the reader's computer.)

■ COMMON FONTS ■

For the text part of your Web page, you'll want to use fonts that are available on most computers. The most common fonts are Times, Arial or Helvetica, and Courier: those four will show up in some form on most computers. Beyond those four, however, things get a little more complicated. Windows and Macintosh have different default fonts. (Windows default fonts are Arial, Book Antiqua, Bookman Old Style, Century Gothic, Century Schoolbook, Courier, Courier New, Garamond, MS Dialog, MS LineDraw, MS Serif, MS Sans Serif, MS System, Times New Roman, and Verdana. Macintosh default fonts are Avant Garde, Bookman, Chicago, Courier, Geneva, Helvetica, Monaco, New Century Schoolbook, New York, Palatino, and Times.) Many Macintosh users have picked up additional Windows fonts through Microsoft programs such as Internet Explorer; however, few Windows owners will have Macintosh fonts such as Monaco. Microsoft has designed two fonts especially for computer monitors: Verdana (sans serif) and Georgia (serif). They're available for free download at http://www.microsoft.com/truetype/. However, because you can never be sure that a reader has any fonts beyond the standard four, you should be careful to include one of those four whenever you designate a font in CSS. If a special font is necessary for something like a title or banner, it's still best simply to create it in a program such as Photoshop and save it as a graphic. It will add some size to your page, but at least you can be sure that the page will look the way you intended.

■ TRY IT

Write the HTML and text for a page using an internal style sheet. Create a class and specify a font-family for it. Remember: Even though you're using CSS, you still need to include the basic template: <HTML></HTML>, <HEAD></HEAD>, <TITLE></TITLE>, <BODY></BODY>

If you think it's appropriate to your design, try changing the font on your text-based or image-based site.

Font-Size

You can also set the size of your font more precisely in CSS than with HTML. You have a variety of options for the value of the font-size property:

- Length designations: points (pt), pixels (px), inches (in), centimeters (cm), or ems
- Absolute expressions: xx-small, x-small, small, medium, large, x-large, xx-large
- Relative expressions: smaller or larger
- A percentage that indicates how much larger or smaller you want the font than its parent element.

■ THE MYSTERIOUS EM ■

You're probably familiar with such measurements as points or pixels, but you may be wondering what the em measurement is. Ems are length units equal to the width of the letter *m* in whatever font you're using. Thus the size of the em depends on the size of the font: 1 em would be the default font size; 2 em would be twice as big; and so on. Many Web designers prefer to use ems as measurements because their size is roughly the same on all monitors, unlike points or pixels, which can vary widely from one computer to another. However, the em measurement doesn't work in all browser versions. If you decide to use it, be sure to check the results.

Both percentages and relative expressions refer to the size of what's called the parent element. Most text elements (i.e., paragraphs, headings, etc.) have preset sizes in HTML; you can use smaller or larger or percentages to indicate how you want your text to look in comparison to those preset sizes. So, for example, if you define H1 as 80%, it will be 80 percent of the size the default H1 is in HTML. Using some of these length designations will still mean that type looks different on Windows and Macintosh monitors: Windows monitors have 96 pixels per inch and Macintosh uses 72 pixels per inch. However, if you use percentages or ems, you'll get sizes that are roughly the same: 100% means full size on both monitors and the width of the letter *m* will be roughly the same on both. If you put these first two properties together, your CSS rule would look like this:

```
H2{font-family:verdana,arial,helvetica,sans-
serif;font-size:12pt;}
```

▄ TRY IT

Use the page you created for the previous exercise. Add font-size to your class.

Font-Style–Font-Weight

To designate italic type, you can use the font-style property. Your choices for values are

- italic
- normal (removes all style designations)

You can use several levels of bold, using the font-weight property. Your choices for values are

- bold
- bolder *or* lighter (sets weight bolder or lighter than parent element)
- values from 100 to 900, in increments of 100
- normal (removes all weight designations, including default boldface for elements like headings)

If we add these properties to the rule, it looks like this:

```
H2{font-family:verdana,arial,helvetica,sans-
serif;font-size:12pt; font-style:italic;font-
weight:normal;}
```

In a browser it would look like Figure 9.1.

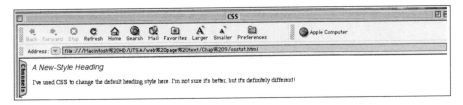

FIGURE 9.1   Font Variations

■ TRY IT

Use the page you created for the previous exercise. Add `font-style` and `font-weight` to your class.

Font

You can set more than one font property in one rule instead of breaking it up into `font-family`, `font-size`, and so on. To set all of these properties at once, use `font` as the property. However, you have to list all of the properties in a particular order (separated by spaces):

- Begin with the `font-style` and `font-weight` values (e.g., italic, bold). These values aren't required, but if you use them, they come first. If you don't use them, you'll get the default values (i.e., none).
- After `font-style` and `font-weight` comes `font-size`, which is required. If you don't specify a font size, the style won't work.
- End with the `font-family` value, which is also required. Separate the font names with commas.

You can also include a `line-height` value within the `font` property if you want to (see the explanation of `line-height` that follows). Put a back-slash after the `font-size` value and write the `line-height` value.

If you set all of these values for font (including `line-height`), the result looks like this:

```
<HEAD>
        <TITLE>CSS Styles</TITLE>
<STYLE TYPE="text/css">
<!--
P{font:italic bold 12pt/14pt verdana,arial,sans-
serif;}
-->
```

```
</STYLE>
</HEAD>
```

Setting Multiple Font Properties

1. Type the selector for the rule followed by a bracket (e.g., `H2{`).
2. Type `font` and a colon (e.g., `H2{font:`).
3. Type these values for the font, ending with a bracket: `font-style` [optional] `font-weight` [optional] `font-size` [required]/`line-height` [optional] `font-family` [required] (e.g., `H2{font:italic bold 10pt/14pt arial,helvetica,sans-serif;}`).

Note: You separate the values with spaces rather than semicolons with this property; however, if you're specifying more than one font family (as you should), separate them with commas.

■ TRY IT

Use the class you created in the last three exercises. Convert the font properties you entered (i.e., `font-family`, `font-size`, `font-style`, `font-weight`) to the `font` property.

Line-Height

CSS allows you to increase the space between the lines of your text so that you can avoid the dense, text-heavy look of some HTML pages. The property is `line-height` and, as with `font-size`, you have some choices for values:

- A length unit: points (`pt`), pixels (`px`), inches (`in`), centimeters (`cm`), or ems.
- A number that's multiplied by the font size to get spacing; if you wanted double spacing, you'd use 2, single spacing 1, and so on. Usually `1.5` to 2 is enough space.
- A percentage that refers to the font size. (For example, 100% would mean a space equal to the height of a line of text or single space; 200% would mean double space; and so on.)

You can define `line-height` separately or as part of the `font` property. As a separate property, it looks like this:

```
.intro{font-family:georgia,times,serif;font-
size:12pt;line-height:2;}
```

To make `line-height` part of the `font` property, put a slash after the font size and add the line height. It looks like this:

```
.intro{font:12pt/14pt georgia,times,serif;}
```

Figure 9.2 shows what it looks like in a browser. The class "intro" is associated with the first paragraph.)

■ TRY IT

Add a `line-height` value to the style sheet your wrote in the last exercises. You can add it to your `font` properties if you want.

Text-Align

You can align text in CSS as you could in HTML, but you can also use full justification, which wasn't available in HTML. Your choices are

- `left` for left justification
- `right` for right justification
- `center` for centered text
- `justify` to align text on the left and right margins

Be careful about using full justification on your Web pages; spaces will be added between your words to make the lines of text stretch to the margins of your page, and the result may look strange, depending on how much text you have.

The CSS rule looks like this:

```
.straight{text-align:center;font-family:
verdana,helvetica,sans-serif;font-size:12pt;
line-height:2;}
```

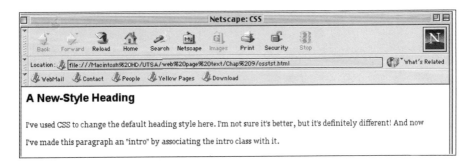

FIGURE 9.2 Line Spacing

In a browser, it looks like Figure 9.3.

Text-Indent

Many Web designers have lamented the fact that in previous versions of HTML the default paragraph style for Web pages didn't include paragraph indentation; most people find indented paragraphs easier to use and better looking than paragraphs separated by spacing. CSS allows you to create indented paragraphs; you can use two methods to measure your indentation:

- A length measurement, such as inches (`in`), centimeters (`cm`), pixels (`px`), or `ems`.
- A percentage, indenting the paragraph proportional to the total paragraph width. (For example, if you wrote `text-indent:10%`, you'd be setting up a paragraph indent that was 10 percent of the paragraph width.)

The code looks like this:

```
P{text-indent:10%;}
```

Figure 9.4 illustrates how it looks in a browser. (Notice that all three paragraphs are indented, although the first two use different classes; that's because the property is set to apply to all paragraphs, no matter what class, by using the P tag (e.g., `P{text-indent:10%):`).

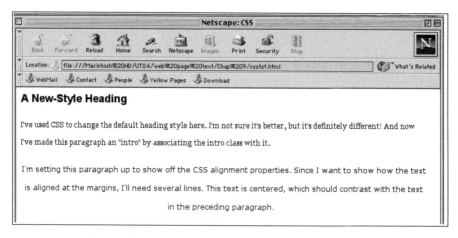

FIGURE 9.3   Centered Text

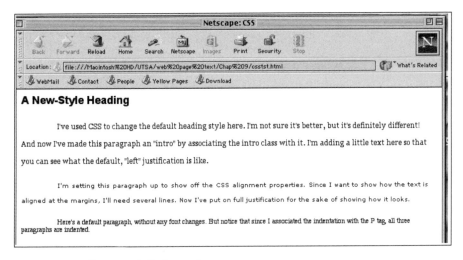

FIGURE 9.4   Paragraph Indentation

Margins

Thanks to CSS, you can now include real margins on your pages, without having to use layout tables to set them up. You have several choices for setting your margins; you can use

- A length measurement, such as pixels (**px**), inches (**in**), centimeters (**cm**), or **ems**
- A percentage (The margin width will be a percent of the page's total width.)
- **auto**, which will set the margins to the default in the browser

You can place margins on each page by using **BODY** as your selector. In that case, the code will look like this:

```
BODY{margin:20px;}
```

In a browser, it looks like Figure 9.5.

BODY{margin:20px;} will give you a margin of twenty pixels around the entire page. You can also set individual margins using **margin-left**, **margin-right**, **margin-top**, *or* **margin-bottom**. If I wanted to set only the left margin for the body text, I could write my code like this:

```
BODY{margin-left:20px}
```

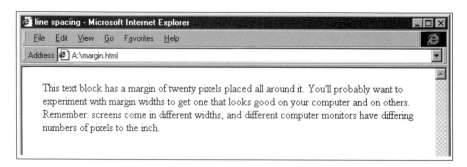

FIGURE 9.5   Margins

You can also set margins around individual elements on the page. For example, if you wanted your first-level headers to be placed closer to the left margin, you could set the H1 margins separate from your paragraph margins. In that case, the definitions would look like this:

```
P{margin-left:40px;font:12pt georgia,times,
serif;}
H1{margin-left:10px;font:18pt helvetica,arial,
verdana,sans-serif;}
```

Figure 9.6 shows what you'd see in a browser.

■ TRY IT

Write the text and HTML, including STYLE section, for a page that uses these properties: text-align, text-indent, margin. Remember: Even though you're using CSS, you still need to include the basic template: <HTML></HTML>, <HEAD></HEAD>, <TITLE></TITLE>, <BODY></BODY>

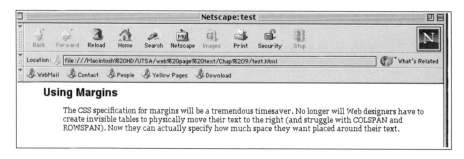

FIGURE 9.6   Element Margins

If these text styles are appropriate to your text-based or image-based site, try adding them.

Color–Background-Color

You can set the color of any element in CSS by using the `color` property. To make all your level-one headings red, you'd write:

```
H1{font:14pt verdana,helvetica,sans-serif;color:
red;}
```

Setting text color is quite similar in HTML and CSS, but the CSS method for setting background colors offers you a lot more flexibility. With CSS, you can utilize different background colors for different areas on a page, while using another background color or a background tile for the page as a whole. So if you want to highlight a particular text block, you can use a different background color to do it. A highlighted section would look like this:

```
.hilite{font:12pt/14pt verdana,helvetica,sans-
serif;color:red;background:yellow}
```

Figure 9.7 illustrates what it looks like in a browser. (The "hilite" class is associated with a DIV.)

■ TRY IT

Write the HTML and text, including a **STYLE** section, for a page that uses the **background-color** property. Remember: Even though you're

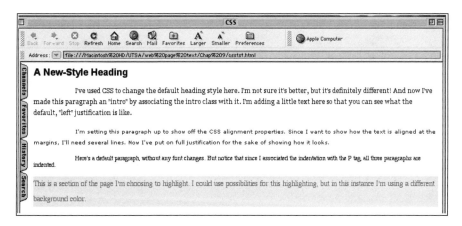

FIGURE 9.7 Background Color

using CSS, you still need to include the basic template: `<HTML></HTML>`, `<HEAD></HEAD>`, `<TITLE></TITLE>`, `<BODY></BODY>`

If this `background-color` property would be appropriate for your text-based or image-based site, try adding it.

Background-Image

You can also use a tiling graphic behind a particular part of a page; so, for example, you could have all of your level-one headings placed on a textured background to highlight them. The property is `background-image` and the code would look like this:

```
H1{background-image:url(background.gif)}
```

Instead of `background.gif`, you'll insert the name of the graphic you're using for your background tile (along with the path, if you've placed your graphic in a directory, that is, `url(images/background.gif)`). In a browser, it looks like Figure 9.8.

■ TRY IT

Write the HTML and text, including a `STYLE` section, for a page that uses the `background-image` property. Remember: Even though you're using CSS, you still need to include the basic template: `<HTML></HTML>`, `<HEAD></HEAD>`, `<TITLE></TITLE>`, `<BODY></BODY>`

Use the `background-image` property on your text-based or image-based site if it's appropriate for your design.

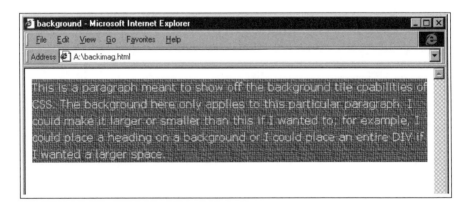

FIGURE 9.8 Background Image

Background-Repetition

Besides letting you insert background images, CSS will also let you determine how you want a background image to repeat. That means you no longer have to worry about an image repeating horizontally in a super-wide browser window. It also means you won't have to use the narrow, very long strip background tile to get a vertical stripe running down the left margin of your page. You can simply choose a background color, such as white, and then insert a background tile that will repeat vertically down the side of the page without repeating anywhere else.

The property here is **background-repeat** and you have several choices for values:

- **repeat**. This value tells the browser to repeat the tile throughout the background, both horizontally and vertically, as it does now.
- **repeat-x**. This value tells the browser to repeat the tile horizontally but not vertically. The graphic will be repeated in one straight, horizontal line.
- **repeat-y**. This value tells the browser to repeat the tile vertically but not horizontally. The graphic will be repeated in a straight vertical line down the left side of the browser window.
- **no-repeat**. This value tells the browser not to repeat the background tile: it will appear only once.

The code will look like this:

```
BODY{background-image:url(background:gif);
background-repeat:repeat-y;)
```

Notice that you must begin by specifying the background image you want to use (including the path, if necessary); then you can determine how that image will show up in the browser. Using this code to create a vertical bar down the left side of the page results in the browser window in Figure 9.9.

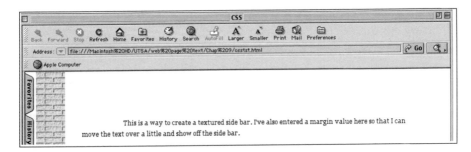

FIGURE 9.9 Vertically Repeating Background

■ TRY IT

Write the HTML and text, including a STYLE section, for a page that uses the background-image and background-repeat properties. Remember: Even though you're using CSS, you still need to include the basic template: <HTML></HTML>, <HEAD></HEAD>, <TITLE></TITLE>, <BODY></BODY>

If it's appropriate to your text-based or image-based site, add the background-repeat property.

Background

You can set all the background values at the same time, as you did with the font values. The property is background, and you can set the values in any order. The code will look like this:

```
BODY{background: url(image.gif) #FFFFFF repeat-y;}
```

■ TRY IT

Write the HTML and text, including a STYLE section, for a page that uses the background property. Remember: Even though you're using CSS, you still need to include the basic template: <HTML></HTML>, <HEAD></HEAD>, <TITLE></TITLE>, <BODY></BODY>

■ CREATING A SIDE BAR OR A TITLE BAR ■

You can use the background-repeat property to create graphic bars that run down the side of your page or that run across the top, behind your title. If you combine this property with others such as background colors and fonts, you can create complex pages without having to create graphics that eat up memory.

Creating a Side Bar or Title Bar

1. Create a thin graphic that's the width you want your side bar to be.
2. For a side bar, set the BODY selector like this: BODY{background: #FFFFFF [insert your choice of background color] url(sidebar.gif) [insert your background image] repeat-y;}
3. For a title bar, set the background of your H1 selector like this: H1{background:#FFFFFF [insert your choice of background color] url(titlebar.gif) [insert your background image] repeat-x;}

Other Text Properties

The properties described in this section are implemented in most version 4.x browsers, as well as a few version 3.x browsers. Other CSS text properties are also available. For example, several properties let you specify a border for your text and specify the border style you want to use; other properties set padding around elements as you did within table cells. However, the other rules are available only in a limited number of browsers at the present time. If you're curious about other rules, consult the reference sites listed at the end of this chapter and test the rules in several browsers to see how widely implemented they are. Several CSS positioning properties also provide you with a way of placing your Web page elements without using layout tables or spacers. Once these properties are fully implemented, you'll be able to place your text and graphics where you want them on the page, without having to worry about how they'll show up on different monitors using different browsers.

Unfortunately, the operative phrase here is *once these properties are fully implemented.* At the moment, position properties are only partially implemented and only in the most advanced browser versions. There are lots of exciting possibilities here—stay tuned for future developments.

HTML, DHTML, AND THE FUTURE

This chapter may seem to present a bewildering number of possibilities, but it's really only a small part of what CSS can do. CSS is only one of many capabilities that Dynamic HTML (DHTML) offers, including the use of JavaScript, CGI scripts, and new ways of interacting with database information. HTML is changing radically as new standards are adopted by the W3C and incorporated by browser manufacturers. It's safe to say that the version of HTML you learned at the beginning of this book will continue to undergo a number of major changes during the next few years.

However, there's one thing to keep in mind: it's still HTML. HTML remains the basis for Web design, although it has been considerably augmented over the years. If you learn the basic workings of HTML, you can put together a credible Web page, even though you may not yet be able to use the latest innovations. Knowledge of HTML becomes a real foundation for the future. Using what you've learned in this book, you can now move on to new developments with a strong sense of how the structure of a Web page is established.

■ WRITING ASSIGNMENT

Try converting one of your earlier Web pages, or even one of your Web sites, to a CSS site. Add a **STYLE** section in the **HEAD** section, including text styles and any background or color effects you'd like to try. Be sure to check your pages in both major browsers; if you can, try checking them in earlier browser versions as well, so that you can see the differences in browser implementations of CSS.

CSS PROPERTIES USED IN CHAPTER 9

CSS PROPERTY	VALUES	EFFECTS
`font-family:`	Font names, separated by commas, ending with a generic designation (i.e., `serif`, `sans-serif`, `monospace`, `cursive`, or `fantasy`)	Designates preferred font styles; browser will check for each name in list and will use the first match it finds
`font-size:`	Length designations (`pt`, `px`, `in`, `cm`, or `em`) or absolute designations (`xx-small`, `x-small`, `small`, `medium`, `large`, `x-large`, `xx-large`) or relative designations (`smaller` or `larger`) or percentages	Indicates the preferred size for the font
`font-style:`	`italic` or `normal`	Indicates style in which font is to be displayed
`font-weight:`	`bold` or `bolder`/`lighter` or values from 100 to 900 (in increments of 100) or `normal`	Indicates degree of boldness in which font is to be displayed
`font:`	`font-style font-weight font-size/line-height font-family`	Establishes five values for fonts in the same property
`line-height:`	Length designations (`pt`, `px`, `in`, `cm`, or `em`) or a number multiplied by the font size (e.g., 2 for double spacing, 1 for single spacing) or percentages	Sets the space between lines of text
`text-align:`	`left` or `right` or `center` or `justify`	Sets text justification for page
`text-indent:`	Length designations (`px`, `in`, `cm`, or `em`) or percentages	Sets paragraph indentation
`margin:`	Length designations (`px`, `in`, `cm`, or `em`) or percentages or `auto`	Sets margins for an entire page
`margin-left:`, `-right:`, `-top:`, or `-bottom:`	Length designations (`px`, `in`, `cm`, or `em`) or percentages or `auto`	Sets individual margins around page or around elements

CSS PROPERTY	VALUES	EFFECTS
`color:`	A hexadecimal color number (preceded by #) or a predefined color name	Sets text color
`background-color:`	A hexadecimal color number (preceded by #) or a predefined color name	Sets background color for an element
`background-image:`	`url(background.gif*)`	Sets background tile for an element
`background-repeat:`	`repeat` or `repeat-x` or `repeat-y` or `no-repeat`	Indicates how a background image is to be repeated; that is, horizontally (`repeat-x`), vertically (`repeat-y`), both (`repeat`), or neither (`no-repeat`)
`background:`	`url(background.gif*)` `colorname**` `repeatname***`	Sets three background values at the same time

*Insert the name (and path, if necessary) of the graphic to be used for a background tile.
**Insert the name of the background color (or the hexadecimal number, preceded by #).
***Insert repeat designation (i.e., `repeat-x`, `repeat-y`, etc.).

TAGS USED IN CHAPTER 9

OPENING TAG	CLOSING TAG	EFFECT
`<LINK REL=stylesheet TYPE="text/css" HREF="filename.css"*>`	None	Links page to external CSS style sheet
`<STYLE TYPE="text/css">`	`</STYLE>`	Creates internal style sheet in HEAD section
`<Selector** STYLE="property:value">`	`</Selector**>`	Creates style within an HTML tag
`CLASS="classname"***`	None	Designates class to be used with an HTML element (placed within opening element tag)
`ID="IDname"***`	None	Designates ID to be used with an HTML element (placed within opening element tag)

OPENING TAG	CLOSING TAG	EFFECT
<DIV>	</DIV>	Creates a division: a page section with a unique style, separated by a space break above and below
<SPAN>	</SPAN>	Creates a span: a page section with a unique style, not separated from the rest of the page by spaces

*Insert the name of the external style sheet with extension `.css`.
**Insert the symbol for an HTML element to be affected by the style.
***Insert the name of the class or ID assigned to the element.

WEB SITES

CSS Overviews

http://home.netscape.com/computing/webbuilding/studio/feature1999v1n5–3.html
Eric Krock's "Top 10 Questions about Cascading Style Sheets"; some answers to typical questions about CSS

http://www.builder.com/computing/webbuilding/powerbuilder/Authoring/CSSToday/
Joseph Schmuller's "CSS Today"; an overview of CSS

http://www.alistapart.com/stories/fear/fear1.html
Jeffrey Zeldman's "Fear of Style Sheets"; both an appreciation of the potential of style sheets and a catalog of problems (with some "workaround" solutions)

http://www.webreview.com/wr/pub/97/04/11/feature/part3.html
David Siegel's "The Web Is Ruined and I Ruined It: Part II, Style Sheets: The Light at the End of the Tunnel"; an appreciation of style sheets from a major Web designer, along with (inevitably) a description of their current problems

CSS Tutorials

http://netcenterbu.builder.com/Authoring/CSS/
"Using Cascading Style Sheets"; a brief, helpful tutorial in CSS basics from CNET

http://www.htmlgoodies.earthweb.com/beyond/css.html
Joe Burns's tutorial on CSS, part of his HTML Goodies tutorial series

http://www.wdvl.com/Authoring/Style/Sheets/Tutorial.html
  Alan Richmond's "Introduction to Style Sheets"; the Web Developer's Virtual Library CSS tutorial

http://www.zdnet.com/devhead/stories/articles/0,4413,1600436,00.html
  "CSS"; part of ZDNet's "32 Ways to Build a Better Web" series

CSS Reference

http://netcenterbu.builder.com/Authoring/CSS/table.html
  Table of properties and values

http://www.projectcool.com/developer/cssref/index.html
  A reference sheet for CSS at Project Cool, a consortium of designers

http://www.wdvl.com/Authoring/Style/Sheets/WDVL.html
  The style sheet used at the Web Developer's Virtual Library site; an example of real-life CSS

http://www.webreview.com/wr/pub/guides/style/mastergrid.html
  Extensive table comparing browser compliance on aspects of CSS

http://www.webstandards.org/
  The Web Standards Project: a consortium of Web designers and writers trying to promote the adoption of common standards across all browsers; a good source of current news about browser compliance.

http://www.w3.org/Style/CSS/Test/
  CSS Test Suite from the W3C; you can use it to test your CSS pages on their compliance with CSS standards.

Creating Frames

USING FRAMES

Frames allow you to divide your page into multiple sections, each of which can be controlled independently. Each of these sections is actually an independent HTML file, and all the frames are pulled together into a master HTML file called a *frameset*. The frameset sets the basic layout of the page: which part will contain which file. When users open a Web page with frames, they actually see the frameset page.

FRAMESETS

Your frameset divides the browser window into separate panes: each pane contains different information, supplied from a different HTML file. The frameset is placed after the **HEAD** section of the page, taking the place of the **BODY** section.

The frameset page contains the basic layout that relates these panes to each other. You indicate whether the page is to be divided horizontally (i.e., into rows), vertically (i.e., into columns), or both. You also indicate how wide you want those rows or columns to be by inserting either a number of pixels or a percent that indicates how much of the window the row or column is to occupy. You can also insert a * to indicate you want this particular row or column to take up all the space left over in the window after the other rows or columns have been filled. So, if you wanted a top row with a banner graphic that was twenty pixels high and a bottom row with a menu that was also twenty pixels high, along with a middle section that would fill up however much space was left after those two rows had been completed, you'd write `FRAMESET ROWS="20, *, 20"`.

You also include names for each of the frames. These names serve as references that you'll use when you target a particular HTML file to appear in a particular frame. Because the frame names are targets, it's best to use names that will remind you of what's supposed to go in a particular frame. (For example, if you're going to put a navigation bar in one frame, you can name that frame navbar.)

As with tables, it's sometimes easier to begin by sketching your frameset layout on paper so that you can see how you want it to be divided. Then you'll have a better idea of how to set it up.

Let's begin by creating a one-column frameset with horizontal rows.

Creating a Frameset with Rows

1. After </HEAD>, type <FRAMESET
2. Type ROWS="x (For x, insert the height of the first row: either a percentage, a number of pixels, or *, which indicates variable height dependent on the height of the other rows.)
3. Type ,y (the height of the second row).
4. Repeat Step 3 for any additional rows.
5. Type ">
6. Type <FRAME NAME="name" (For name, insert the name of the frame in the first row; i.e., at the top of the screen.)
7. Type SRC= "url.html"> (For url.html, insert the URL for the page to be displayed in this frame.)
8. Repeat Steps 6 and 7 for each row, moving from top to bottom.
9. Type </FRAMESET> and </HTML> to complete the frameset.

Code for a frameset with three rows would look something like this:

```
<HTML>
<HEAD>
<TITLE>Three-Row Frameset</TITLE>
</HEAD>
<FRAMESET ROWS="20, *, 20">
<FRAME NAME="banner" SRC="banner.html">
<FRAME NAME="welcome" SRC="welcome.html">
<FRAME NAME="menu" SRC="menu.html">
</FRAMESET>
</HTML>
```

The code for setting up a one-row frameset with three columns would be similar.

Creating a Frameset with Columns

1. After </HEAD>, type <FRAMESET
2. Type COLS="a,b"> (For a,b, insert the width of the columns: either a percentage, a number of pixels, or *, which indicates variable width dependent on the width of the other columns.)
3. Type <FRAME NAME="name" (For name, insert the name of the frame in the first column; i.e., at the left of the screen.)
4. Type SRC="url.html" (For url.html, insert the URL for the page to be displayed in the first column.)

5. Repeat Steps 3 and 4 for each additional column, moving from left to right.

6. Type </FRAMESET> to complete the frameset

The code would look like this:

```
<HTML>
<HEAD>
<TITLE>Two-Column Frameset</TITLE>
</HEAD>
<FRAMESET COLS="20, *">
<FRAME NAME="menu" SRC="menu.html">
<FRAME NAME="welcome" SRC="welcome.html">
</FRAMESET>
</HTML>
```

Finally, you can create a frameset with both columns and rows; that is, one that's divided both horizontally and vertically. This kind of frameset can become complex; remember to sketch out the layout first before you start writing the code. These instructions will create a frameset with three rows, the center one of which will be divided into two columns. Notice that here you actually have one frameset (the two columns) nested inside another frameset (the three rows).

Creating a Frameset with Columns and Rows

1. After </HEAD>, type <FRAMESET

2. Type ROWS="x,*,y"> (Insert the height of the three rows.)

3. For row 1, type <FRAME NAME="name" SRC="url.html">

4. For row 2, type <FRAMESET COLS="a,b"> (Insert the width of the two columns contained in row 2.)

5. Type <FRAME NAME="name" SRC="url.html"> for the first column.

6. Repeat Step 5 for the second column.

7. Type </FRAMESET> to conclude the inner frameset (the columns).

8. Type <FRAME NAME="name" SRC="url.html"> and insert the name of row 3.

9. Type </FRAMESET> to conclude the outer frameset (the rows).

The code for this frameset would look something like this:

```
<HTML>
<HEAD>
<TITLE>Column and Row Frameset</TITLE>
</HEAD>
```

```
<FRAMESET ROWS="20, *, 20">
<FRAME NAME="banner" SRC="banner.html">
<FRAMESET COLS="20, *">
<FRAME NAME="dates" SRC="dates.html">
<FRAME NAME="welcome" SRC="welcome.html">
</FRAMESET>
<FRAME NAME="menu" SRC="menu.html">
</FRAMESET>
```

LINKING FRAMES

Once you've created the frameset to contain your frames, you'll need to create the frames to go into it. The first frames to show up in a frameset will be those you specify in the SRC codes you wrote in the original frameset page code. But one of the advantages of frames is the ability to open new pages within the individual "panes" of the frameset window.

To open a new frame within the frameset, you need to create a link between the frameset and the new frame. These linked frames are called *targeted links*. To target a link, you must be certain to assign a name to each frame in the FRAME tag; you'll use this name to identify the target window where you want the new frame to open. Use the following instructions to replace the initial content of a frame with another page when the user clicks on a link.

Linking a Page to Open in a Frameset

1. Create a link on the page where the second frame will replace the first, using the standard link format <A HREF="page2.html" (Substitute the URL of your second page for "page2.html".)
2. Type TARGET=name> (Instead of name, insert the name of the frame you used in the FRAME tag in the FRAMESET code.)

This is what the code on the linked page would look like:

```
<HTML>
<HEAD>
      <TITLE>sidebar</TITLE>
</HEAD>
<BODY BGCOLOR="white" Text="red">
<P STYLE="font-family: Verdana, Arial,
Helvetica; font-size: 12pt/14pt">This is the
sidebar part of the second row. It would usually
hold the menu, or the navigation bar. So here are
some targeted links:</P>
<A HREF="frm2.html" TARGET=Calavera>Page 1</A>
<BR>
```

```
<A HREF="frm2b.html" TARGET=Calavera>Page 2</A>
</BODY>
</HTML>
```

The links on this page will open in a frame in the frameset called Calavera. (You can see the complete text for this page and the others in this frameset in Figures A.1 and A.2, pages 205 and 206.)

You can also open a frame in a separate window so that it isn't constrained by the dimensions of the original. You'll want to do this for external links, so that the Web pages you link to will come up at full size.

Linking a File to Open in a Different Page

- Create a link, using the standard `<A HREF="url"`
- Type `TARGET=_blank` to have the link open in a new blank window *or*
- Type `TARGET=_self` to have the link open in the frame that contains the link *or*
- Type `TARGET=_top` to open the link in the browser window independent of the rest of the frameset

IMPROVING THE LOOK OF YOUR FRAMES

You can do several things to your frameset to improve its appearance and to take care of any problems with crowding between frames. First, you can adjust the frames' margins. The default margin for frames is eight pixels on a side, but you can add or delete space if you wish.

Adjusting Frame Margins

1. Within the `<FRAME` tag, type `MARGINWIDTH=x` (Insert the amount of space you want between the frame's left and right edges and the frame's contents, in pixels.)
2. Type `MARGINHEIGHT=y` (Insert the amount of space you want between the edge of the frame's top and bottom and the frame's contents, in pixels.)

You can also remove the borders around frames, so that they have no visible divisions (except for scroll bars).

Eliminating the Frame Border

1. Inside the first `<FRAMESET` tag, type `FRAMEBORDER=0  FRAMESPACING =0 BORDER=0`
2. To eliminate only the vertical borders, inside each `<FRAME` tag type `FRAMEBORDER=0`

3. To eliminate only the horizontal borders, define the columns in the outer frameset and the rows in the inner frameset; type **FRAMEBORDER=0** within each of the **FRAME** tags in the columns.

Finally, you can adjust the scroll bars that usually appear around frames, so that they take up less space in the window. You can hide a frame's scroll bars or show them only when the frame exceeds a certain length or width.

Hiding Scroll Bars

- To *show scroll bars continually,* in the <FRAME tag, type SCROLLING=YES
- To *hide scroll bars continually,* in the <FRAME tag, type SCROLLING=NO
- To *show scroll bars when necessary,* in the <FRAME tag, type SCROLLING =AUTO

Automatic scroll bars are the default setting; if you want scroll bars only when necessary, you need not add anything to the page.

Now, here are some examples of code and the frame pages they produce. This is the frameset code:

```
<HTML>
<HEAD>
        <TITLE>frames 1</TITLE>
</HEAD>
        <FRAMESET ROWS="15%,*,10%" FRAMEBORDER=
"0" FRAMESPACING="0" BORDER="0">
<FRAME SRC="frm1.html" NAME="first" SCROLLING
="no">
<FRAMESET COLS="15%,*"><FRAME NAME="sidebar"
SRC="frm2a.html">
<FRAME NAME= "Calavera" SRC="frm2.html">
</FRAMESET>
<FRAME SRC="frm3.html" NAME="Third" SCROLLING=
"auto">
</FRAMESET>
</HTML>
```

This is the text that will appear at the top of the page in the frame named first.

```
<HTML>
<HEAD>
        <TITLE>first</TITLE>
```

```
</HEAD>
<BODY BGCOLOR="white">
<IMG SRC="images/frames.gif">
</BODY>
</HTML>
```

This text will appear in the narrow column to the left, the frame named sidebar.

```
<HTML>
<HEAD>
        <TITLE>sidebar</TITLE>
</HEAD>
<BODY BGCOLOR="white" Text="red">
<P STYLE="font-family: Verdana, Arial,
Helvetica; font-size: 12pt/14pt">This is
the sidebar part of the second row. It
would usually hold the menu, or the
navigation bar. So here are some targeted
links:</P>
<A HREF="frm2.html" TARGET=Calavera>Page 1</A>
<BR>
<A HREF="frm2b.html" TARGET=Calavera>Page 2</A>
</BODY>
</HTML>
```

This text will appear in the main part of the page, the frame named Calavera.

```
<HTML>
<HEAD>
        <TITLE>Second</TITLE>
</HEAD>
<BODY BGCOLOR="White" TEXT="red">
<IMG SRC="images/calavera.gif" ALIGN="left">
<P STYLE="font-family: Verdana, Arial,
Helvetica; font-size: 12pt/14pt"> This is the
content of the frame that will take up all the
space left over when the other frames have been
filled. It contains a graphic and this text
section, and it will scroll so that the entire
graphic (which is rather large) can appear on the
screen at the same time. I've made all of these
frames with the same background color. But when I
```

```
change to the other page, the background color
will change too.</P>
</BODY>
</HTML>
```

This text will appear at the bottom of the page in the frame named third.

```
<HTML>
<HEAD>
        <TITLE>Third</TITLE>
</HEAD>
<BODY BGCOLOR="white" TEXT="red">
<P STYLE="font-family: Verdana, Arial,
Helvetica; font-size: 12pt/14pt">And this is the
content of the third frame, which should be the
same size as frame number one. This is a fairly
complex layout, more complex than I would usually
have in a frameset.</P>
</BODY>
</HTML>
```

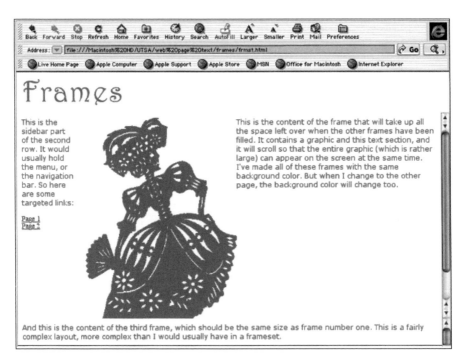

FIGURE A.1   Frameset

Finally, this text will appear when the link to page 2 is clicked on the sidebar menu. Notice that the page is targeted to frame Calavera.

```
<HTML>
<HEAD>
      <TITLE>calavera</TITLE>
</HEAD>
<BODY BGCOLOR="yellow" TEXT="red"><BASEFONT
SIZE="4">
<IMG SRC="images/calavera2.gif" ALIGN="left">
<P STYLE="font-family: Verdana, Arial,
Helvetica; font-size: 12pt/14pt"> This is the
linked page that will come up when the targeted
link is clicked.</P>
</BODY>
</HTML>
```

In a browser the opening frameset looks like Figure A.1.

The linked page looks like Figure A.2.

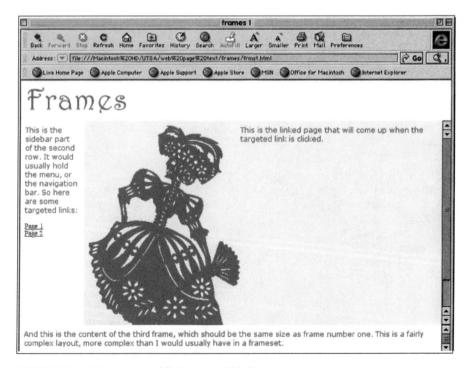

FIGURE A.2 Frameset with Targeted Link

TAGS USED IN THIS APPENDIX

OPENING HTML TAG	CLOSING HTML TAG	EFFECT
`<FRAMESET>`	`</FRAMESET>`	Creates the container for the frames
`<FRAMESET ROWS="x, y*">`	`</FRAMESET>`	Establishes the number of rows and their dimensions in a horizontal frameset; for **x** and **y** insert a number of pixels, a percentage, or *
`<FRAME NAME= "name"** SRC= "url"***>`	None	Gives the name of each frame in the set, as well the URL for the page to be inserted in the frame
`<FRAMESET COLS="x, y"*>`	`</FRAMESET>`	Establishes the number of columns and their dimensions in a vertical frameset
`<FRAME NAME= "name"** SRC= "url"*** MARGINWIDTH="x"* MARGINHEIGHT="y"*>`	None	Inserts a number of pixels between the frame's left and right edges and its content (`MARGINWIDTH`) and between the frame's top and bottom and its contents (`MARGINHEIGHT`)
`<FRAME NAME= "name"** SRC="url"*** SCROLLING="yes"` *or* `"no"` *or* `"auto">`	None	Shows scroll bars continually (`"yes"`), never (`"no"`), or as needed (`"auto"`)
`<FRAMESET FRAMEBORDER="0" FRAMESPACING="0" BORDER="0">`	`</FRAMESET>`	Turns off all visual divisions within the frameset except for scroll bars
`<FRAME NAME= "name"** SRC="url"*** FRAMEBORDER="0">`	None	Turns off only vertical frame borders
`<A HREF="url" TARGET="name"` *or* `TARGET="_blank"` *or* `TARGET="_self"` *or* `TARGET="_top">`	`</A>`	Replaces the initial contents of a frame with another page; for **name** substitute the name of the location frame used in the initial frameset definition to open the page in the same frame; or use `_blank` to open the link in a new window or use `_self` to open the frame in the window containing the link or use `_top` to open the link in the browser window independent of the frameset

GLOSSARY

■ ■ ■ ■ ■ ▬▬▬▬▬▬▬▬▬▬▬▬▬▬▬▬▬▬▬▬▬▬▬▬▬

Note: Italicized terms within definitions are also defined in this glossary.

Absolute Links   A link that includes every part of the *URL* address for a Web page (including protocol and *domain* name), so that the browser can locate exactly that page. Used primarily for links off a Web site.

Anchor Links (Fragments)   A link that connects information on a single page. Anchor links don't use complete *URL*s; instead they refer to an anchor, a name inserted in the code for the page.

Attribute   In *HTML,* words or symbols within *tags* that affect the appearance or behavior of the elements on the page; for example, size, color, align, and the like.

BMP   Bit-mapped; the graphic file format used by Microsoft Windows.

Browser   An application program that uses *HTML* to read and display Web pages; the two most widely used browsers are Netscape Navigator and Microsoft Internet Explorer.

Cell Padding   In *HTML* tables, cell padding provides space around the data placed within the table cell.

Cell Spacing   In *HTML* tables, cell spacing provides space between the table cells.

CGI   Common Gateway Interface; a means of transferring data between a Web server and a specific program; frequently used to provide dynamic interaction between users and Web pages, for example, to process data from forms submitted by Web site visitors.

Class   In *CSS,* a special type of *selector* that can be applied to certain elements; classes allow you to create custom *HTML tags.*

Clip Art   Copyright-free art available either in print form or electronically on CD-ROM or downloaded from the Web. When you purchase clip art, you purchase the right to use it in publications or on your Web pages. However, some clip-art collections place restrictions on the number of images you can use within a page.

CSS   Cascading Style Sheets. A new kind of *HTML* coding that provides advanced text and page design features in a format similar to the style sheets found in word processor and page-layout programs.

Dead-End Pages   Web pages that have no links, providing users no way out.

Display Fonts   Type *fonts* designed to be used for stylistic effects in large sizes (e.g., Anna, Bergell, Homely, Umber, etc.).

Dithering   The way in which Web *browsers* that support only 255 colors approximate other colors. The browser will replace the nonsupported color with a mixture of two or more supported colors; the result may not truly reproduce the original color and may be somewhat grainy on the screen.

DIV   In *CSS,* a tag that creates a division on the page, separated by a break above and below, allowing you to create separate page sections to which you can apply unique styles.

207

Domain   Part of a *URL* that indicates the address of a given Web site (e.g., `www.yahoo.com`; `www.excite.com`; `www.northernlight.com`), indicating both the *server* name and the purpose of the site. (`.com` indicates a commercial purpose.)

Em   Length unit equal to the width of the letter *m* in whatever *font* is being used.

EPS   Encapsulated PostScript; graphics file format used by the PostScript language.

Extension   A series of letters at the end of a file name, usually separated from the file name by a period, for example, `soccer1.html`. The extension tells the *browser* what kind of information is stored in the file, so `.html` indicates that the information is written in *HTML*.

Font   A complete set of letters in a given type style (e.g., Times, Helvetica, Futura, etc.).

Frames   A way to divide Web pages into multiple, independently controlled sections, each using an independent HTML file.

Frameset   A master *HTML* file on a *frames* page that includes the basic layout of the page, indicating which part of the page includes which HTML file.

GIF   Graphics Interchange Format, a graphics format developed by Unisys Corporation. GIFs are highly compressible and can squeeze down to small sizes. They handle flat colors relatively well and are usually used for drawings and line art.

Hexadecimal Color   Codes that convert RGB colors (red, green, and blue—the standard color screen for computer monitors) to mathematical equivalents that Web *browsers* can understand. Hexadecimal color codes always begin with # and always have six letters and/or numbers.

Home Page   The main page of a Web site, which frequently serves as a table of contents to the site's other pages.

HTML   HyperText Markup Language. The coding language used to create tags for the Web.

http   HyperText Transfer Protocol. The Web protocol that defines how information is formatted and transmitted from server to server.

ID   In *CSS,* a special type of *selector* that you can apply to certain elements. However, unlike *classes,* IDs can be applied to only a single element on a page; thus they are usually reserved for applications such as writing scripts or using *CSS* positioning rules.

JPEG   A graphic format developed by the Joint Photographic Experts Group and used most frequently for photographs on the Web. JPEGs do not handle flat colors well and are not usually used for line art or drawings.

Leading   The space between lines of type, measured in points (pronounced "ledding").

Logical Formatting   *HTML tags* that indicate that a text should be changed but that allow the browser to dictate what that change should look like (e.g., italicized in some browsers, but underlined in others).

Mount (Upload)   To transmit Web pages to an Internet *server.*

Navigation Bar   A linked list of major pages branching off from a *home page* or subordinate page at a Web site.

Path   A route through a series of files or file systems to find a particular file; *URLs* can contain path designations. (For example, `/images/idea.gif` indicates that the browser is to find the file `idea.gif` in the directory `images`; the path is through the directory `images` to the file `idea.gif`.)

PDF Portable Document Format; a document file format developed by Adobe Systems that allows you to send formatted documents, while retaining their original format. PDF files are created using Adobe Acrobat and must be read using Adobe Acrobat Reader.

Physical Formatting *HTML tags* that dictate precisely what the text will look like, no matter what the browser defaults are; these tags are unreadable in some browsers.

PICT A file format from Apple computer that can handle both bit-mapped and object-oriented images.

Pixel A single point on the screen of a graphics-reading computer monitor. *Pixel* stands for *picture element;* on color monitors, each pixel is composed of three dots colored red, green, and blue that converge at some point. The quality of the color monitor depends on the number of pixels it can display and also the number of bits used to represent each pixel, ranging from 8-bit systems (256 colors) to 24-bit systems (more than 16 million colors).

PNG Portable Network Graphics, a royalty-free public domain graphic format that supports a variety of advanced features. These graphics are most appropriate for line art and drawings, but are not recognized by browsers earlier than version 4.x.

Property In *CSS,* what's being defined in a style rule. Properties include the way an element looks (e.g., color, size) and the way it acts (e.g., position).

Relative Links Links using Web page *URL*s that don't supply the protocol and *domain* names because the browser is simply moving from one page to another in the same domain using the same protocol. These links are called relative because they're relative to other pages on the same site or in the same directory; usually used for on-site links.

Rule In *CSS,* a formula that defines how *HTML elements* will look and behave when they're viewed through a *browser.*

Sans Serif Without serif. *Fonts* that have no serifs at the ends of letters; for example, Helvetica.

Screen Font A *font* that has been designed especially for use on computer monitors (e.g., Arial, Chicago, Geneva, Georgia, Verdana).

Selector In *CSS,* a word or a set of characters that identifies the style that's being defined. It can be an *HTML tag* (e.g., `P`) or a *class* or *ID selector.*

Serif Short extensions at the ends of letters in some *fonts;* for example, Times.

Server A computer that manages network resources, providing access to files, databases, and so forth. A Web server supports documents formatted in *HTML* and provides access to them for users of the Web.

SPAN In *CSS,* a tag that creates a section on the page, allowing you to create separate page sections to which you can apply unique styles. Unlike the *DIV* tag, SPANs have no spaces showing where they begin and end and can be used to create styles within a larger segment of a page.

Splash Page A page with attention-getting graphics or other effects that opens a Web site (it "makes a splash").

Tag   A command that specifies how part of a Web page should look or behave. For example, `<B>` is a tag that specifies that a word or words should be shown in boldface.

Targeted Links   In *frames,* the link between the *frameset* and an *HTML* file that will open in a particular frame within the set.

Template   A pattern or guide for making pages with a standard format. (For example, the standard *HTML tags* appearing on all HTML pages—`<HTML></HTML>` `<HEAD></HEAD>` `<TITLE></TITLE>` `<BODY></BODY>`—form a Web page template.)

Text-Only File   ASCII code without any of the formatting provided by a word processor. Text-only files are readable on most computers and don't rely on word processing software for translation.

TIFF   Tagged Image File Format. A widely used graphic file format for saving bit-mapped graphics.

URL   Uniform Resource Locator. The address of a file on the Internet.

Value   The definition of the *attribute (HTML)* or the *property (CSS)* in a tag; for example, in the tag `<FONT COLOR="red">`, `red` is the value.

W3C   The World Wide Web Consortium. The governing body of the Web.

WWW   The World Wide Web; a section of the Internet that supports documents coded in *HTML.*

WYSIWYG   What you see is what you get. Applications such as word processors that allow a user to see on the display screen exactly what will appear when the document is printed. Some graphic-interface *HTML* editors are called WYSIWYG because they create *HTML* code while displaying a version of what the code should look like in a Web *browser.*

BIBLIOGRAPHY

Burns, Joe. *HTML Goodies.* Indianapolis: Macmillan, 1999.

Lie, Håkon Wium, and Bert Bos. *Cascading Style Sheets: Designing for the Web.* 2nd ed. Harlow, England: Addison-Wesley, 1999.

Mullet, Kevin, and Darrell Sano. *Designing Visual Interfaces: Communication Oriented Techniques.* Upper Saddle River, NJ: Prentice Hall, 1995.

Nielsen, Jakob. *Designing Web Usability.* Indianapolis: New Riders Publishing, 2000.

Pirouz, Raymond. *Click Here: Web Communication Design.* Indianapolis: New Riders Publishing, 1997.

Siegel, David. *Creating Killer Web Sites.* 2nd ed. Indianapolis: Hayden Books, 1998.

Veen, Jeffrey. *HotWired Style.* San Francisco: Wired Books, 1997.

Weinman, Lynda. *Deconstructing Web Graphics.2.* Indianapolis: New Riders Publishing, 1998.

———. *Designing Web Graphics.3.* Indianapolis: New Riders Publishing, 1999.

———, and William Weinman. *Creative HTML Design.* Indianapolis: New Riders Publishing, 1998.

INDEX